Practical guide to structural robustness and disproportionate collapse in buildings
October 2010

The Institution of Structural Engineers

Constitution of Task Group

Dr A P Mann FREng BSc(Eng) PhD CEng FIStructE MICE (Jacobs) *Chairman*
S J Alexander MA CEng FIStructE FICE MCMI (WSP Group)
J N Carpenter BSc(Eng) CEng FIStructE FICE CFIOSH (SCOSS)
J P Cartz CEng FIStructE (Capita Symonds/Capita Architecture)
Prof M Chryssanthopoulos BSc MS PhD DIC CEng FIStructE FICE (University of Surrey)
G T Harding OBE DIC CEng FIStructE MICE (Consultant)
Dr A E K Jones BEng(Hons) PhD CEng FICE (Arup)
P Kelly BSc(Eng) CEng MIStructE (Treanor Pujol Limited)
G Lewis MEng(Hons) CEng MIStructE MICE (CCB Evolution Ltd)
A Thirumoolan BSc(Hons) CEng MICE FRICS (Wandsworth Borough Council)
J N Tutt MPhil CEng FIStructE (Jenkins and Potter)
T C Cosgrove MSc DIC CEng MIStructE MIEI (SIAC Tetbury Steel)

Secretary to the Task Group

B Chan BSc(Hons) AMIMechE (The Institution of Structural Engineers)

Acknowledgements

Permission to reproduce extracts from British Standards is granted by the British Standards Institution (BSI). British Standards can be obtained in PDF or hard copy formats from the BSI online shop: www.bsigroup.com/Shop or by contacting BSI Customer Services for hard copies only: Tel: +44 (0)20 8996 9001, Email: cservices@bsigroup.com

Colin Jolly, Box 2.5
NHBC, Figure 4.1
Henry Bardsley RPR, Box 4.2
Colin Bailey, Box 4.3
David Longstreath / AP Press Association Images, Box 5.1
Corus Bi-Steel, Figure 8.1, Figure 8.2
CCB Evolution, Figure 9.2, Figure 9.3, Figure 9.10, Figure 9.11
LDSA and LABC, Appendix 1

Published by the Institution of Structural Engineers
International HQ, 11 Upper Belgrave Street, London SW1X 8BH
Telephone: +44(0)20 7235 4535 Fax: +44(0)20 7235 4294
Email: mail@istructe.org, Website: www.istructe.org
First published 2010
ISBN 978-1-906335-17-5

©2010 The Institution of Structural Engineers

Contents

Contents

Tables

Boxes

Glossary

Term / Abbreviation	Definition
Accidental action	Unintended action, usually of short duration but of significant magnitude, which might occur on a given structure during the design working life.
Bridging element	See transfer element.
Building class	The category of building as defined in Table 11 of Approved Document A (Building Regulations: England and Wales) for the purposes of design against disproportionate collapse. BS EN 1991-1-7 Annex A, Table A1 is similar but refers to consequence class.
Disproportionate collapse	A collapse, after an event, which is greater than expected given the magnitude of the initiating event. The level of collapse expected for certain events may be given in regulations, but more often is related to public and professional perception. Even the most robust structure may suffer complete collapse if the event is severe enough, and this would not be considered disproportionate. BS EN 1991-1-7 recognises that total collapse may sometimes be acceptable.
Ductility	The ability of a structure to deform plastically under load without fracture yet still fulfil a load carrying function. In some circumstances, e.g. when considering dynamic loads, the energy absorbed during deformation becomes an important resistance characteristic.
Element removal	An analytical procedure where structural elements are theoretically removed one at a time while the residual structure is checked against specified limits of collapse.
Event or initiating event	In the context of structural robustness, an event can be considered as the occurrence of an accidental action, for example an explosion, an impact, an overload, a fire. Some events will be a combination of other events; deterioration may also play a part in defining an initiating event.
Hazard	A term used in Approved Document A, synonymous with accidental action, having the potential to cause disproportionate damage or collapse.
Hybrid structures	Structures whose elements are made from different materials e.g. precast units on steel or reinforced concrete frames.
Key elements	Selected elements which are designed to withstand a prescribed hazard loading, for example, an applied pressure of 34kN/m^2.
Load path	The complete route via which any applied action (vertical or horizontal) is transmitted through a structure to its foundations through a system of interconnected elements.
Malicious action	Deliberate attempt to cause gross damage to a structure.
Overall responsibility	The principle that for any structure there should be one guiding hand overseeing the structure's overall stability and provisions for robustness. Many structures are an assembly of components which might be adequate on an individual basis but need to be assessed as a whole system as they are frequently interdependent especially from a stability standpoint.
Progressive collapse	The sequential spread of local damage from an initiating event, from element to element, resulting in the collapse of a number of elements. Whilst undesirable, a progressive collapse may not be disproportionate. Hence the term 'progressive collapse' is not necessarily equivalent to 'disproportionate collapse'.
Redundancy	A term used to signify that there are more load paths than strictly necessary to carry the load through the structure (or a part thereof). In structural analysis redundancy is associated with structural indeterminacy, but in the context of robustness the term has a wider meaning and interpretation.
Robustness	A quality in a structure/structural system that describes its ability to accept a certain amount of damage without that structure failing to any great degree. Robustness implies insensitivity to local failure. BS EN 1991-1-7 provides one definition of robustness as "the ability of a structure to withstand events like fire, explosions, impact or the consequences of human error without being damaged to an extent disproportionate to the original cause".
Sensitivity	The concept of a minor change in geometry, assumption, load or resistance (amplitude or direction) having a disproportionate effect on the structure as a whole.
Solidity	A term used in the Workplace (Health, Safety and Welfare) Regulations 1992 without definition. It can be taken to be synonymous with robustness.
Stability	Stability can be sub-divided into 'global stability', 'member stability' and 'local stability'. For all of these, lack of stability implies a gross change of state under increasing loading (e.g. by overturning or buckling). It is necessary to consider stability in the permanent and temporary construction conditions. Many structures are fully restrained and stable once complete, but are quite flimsy in the interim phase.
Ties	Physical tying between elements with the objective of preventing separation.
Transfer element	A beam or slab which carries one or more columns over an increased span where the vertical elements are not aligned from the floor below to the floor above. Its failure will normally imply multiple column failure. A transfer element is the more familiar term for a bridging element.

Foreword

This *Guide* has been prepared by an Institution of Structural Engineers' Task Group responding to a perceived need for more information on structural robustness. That need was identified by the UK Standing Committee on Structural Safety (SCOSS) who determined a lack of confidence amongst designers on the application of Building Regulation requirements. The *Guide's* target audience is engineers charged with designing the bulk of building structures which are relatively low rise, and for example in the UK would be limited to Class 2 in the guidance to the Regulations. The *Guide* does not cover Class 3 structures.

Incorporating robustness in buildings is essential and a stated aim of most regulations and all material codes worldwide. However, robustness is not a commodity readily defined. Hence, as well as providing interpretation and practical guidance on the regulations and material specific construction practice, this *Guide* also contains some generic background on the fundamental attributes of robustness.

The Structural Eurocodes (BS EN 1990 – BS EN 1999) are now available throughout the EU. They will also be adopted in other countries around the world. Although not yet mandatory within the UK, their use is becoming increasingly common, not least because maintenance of existing British Standards ceased on 1 April 2010. Nonetheless, although over time current BS codes will fall out of use, there may be a long period throughout which they will remain in use as a means of demonstrating compliance with Building Regulations. Moreover, many aspects of the BS approach to robustness feature within the Eurocodes (which themselves contain a limited amount of information on accidental damage) largely as a result of the UK's long standing experience with disproportionate collapse which has been a requirement of Building Regulation approval for many years.

This document generally adopts the guidance for achieving robustness as described in Approved Document A (and similar documents for Scotland and Northern Ireland). To cater for the code transition period, the *Guide* defines the formal requirements of the Eurocodes and highlights where these differ from traditional UK practice; thereafter, detailed advice is given on the application of current practice with compliance to British Standards. Where additional guidance from industry is quoted, this is compatible with Eurocode requirements but may be based around British practice. Eventually, the Eurocodes, plus the National Annex, will constitute a very similar approach to British Standard practice and this will be supplemented with NCCI (non contradictory complementary information) to capture any guidance not transferred.

Thanks are due to the significant effort put in by all members of the Task Group.

Allan Mann
Chairman

1   Introduction

Our infrastructure requires a great deal of investment though the demands placed on it over its life remain uncertain. To help cope with such uncertainty, engineers generally consider that attributes like stability, solidity and robustness are 'good things'. Yet because these are abstract qualities, they defy precise codification. The failure of the Ronan Point flats in London in 1968[1.1] highlighted the danger of having non-robust structures and that event changed our attitudes forever. Subsequently, rules were drafted to target the introduction of some measure of robustness into all new buildings. Despite those rules, the concepts remain hazy for many and there is much evidence that practising engineers lack confidence and guidance on how to incorporate robustness and how to comply with regulations.

Since the Ronan Point failure there have been other incidents highlighting the potential vulnerability of structures to severe events. The failure at Oklahoma[1.2] was one and in 2001, the devastating collapse of the World Trade Centre in New York rekindled international attention on building robustness. Perhaps less discussed is the air crash on the Pentagon building on the same day. That event is of interest since it demonstrated the survivability of a well built structure, even to extreme events.

This *Guide* is not about complex rules for preventing progressive collapse in complex buildings. It does not provide guidance on the special case of Class 3 structures in the UK. Rather, the text is targeted at those charged with designing and constructing the smaller, everyday structures, which make up the bulk of our profession's workload.

The *Guide* explains statutory requirements and offers advice on compliance but beyond that, it tries to provide common sense advice on what constitutes robustness; a quality that professional designers ought to incorporate as a matter of prudence. It must be noted that this *Guide* is intended as guidance only and as such cannot be used as a substitute for an Engineer's own professional experience, knowledge, verification and attention.

Whilst recognising the regulations, the Task Group believe that the main thrust of the designer's effort should be on achieving compliance via their professional competence and judgement. The robustness topic does not yet lend itself to academic precision. Indeed, the rules we have now are a pragmatic balance between cost and perceived risk which is nevertheless judged to be reasonably effective in restricting the extent of failure when put to the test. Achieving robustness is partly dependent on a risk perception:
(1) it requires a conscious recognition of initiating events and their likelihood;
(2) it requires a sound understanding of building/ structural performance to assess how these might respond under overload or accident;
(3) it requires judgement about consequences.

A degree of robustness is a sound policy to adopt to prevent economic loss and to limit the risk of harm to the public. Indeed it has been argued that given the current technical state of codes and safety factors, the greatest risks to structures in reality lies not with having inadequate safety margins but with having inadequate robustness.

We all know more or less what is required but mostly by exception:
(1) we don't want structures to fail like a house of cards;
(2) we don't want minor errors to have a disproportionate effect;
(3) we don't want structures to fail to any great degree under accidental loading.

A difficulty lies partly in defining which accidental effects should be considered and thereafter in defining what damage can be tolerated. Because the attribute of robustness is linked to stability, though remains separate from it, (an unstable structure is certainly not robust) the *Guide* makes reference to stability throughout, during both the construction and building complete stages.

When considering robustness in this *Guide*, account has been taken of the varying statutory requirements within the UK. These are outlined in Chapter 3; however in summary form they are:
– Safety legislation applies to Great Britain and Northern Ireland separately although the detailed requirements are the same.
– Building Regulations are governed by separate requirements in England and Wales, Scotland, and Northern Ireland. There is much commonality which is to be expected but some subtle differences do apply. These may change over time and at the time of writing, all official guidance is being reviewed.

The text of the *Guide* is based around the regulations in England and Wales; however if there are relevant variations elsewhere within the UK, these are indicated.

References

1.1   Ministry of Housing and Local Government. *Report of the inquiry into the collapse of flats at Ronan Point, Canning Town.* London: HMSO, 1968

1.2   Corley, W.G. et al. 'The Oklahoma City bombing: summary and recommendations for multihazard mitigation'. *ASCE Journal of Performance of Constructed Facilities*, 12(3), August 1998, pp100-112

2 Concepts of robustness

2.1 Introduction

Structures should be safe. Put simply, the public expects that the possibility of buildings or their parts failing should be so remote as to pose no significant danger to life; the phrase 'as safe as houses' is not in common parlance for nothing. Equally, the public do not expect any significant loss of building asset from everyday events. To a lesser degree, that intolerance extends to rarer events such as vehicle impacts, fire, gas explosions or minor structural alterations and the attribute that imparts those qualities is robustness. The implicit obligation to design structures for robustness is embodied within regulations and codes though the manner of compliance is often not explicit and is left up to good practice and professionalism.

BS EN 1990[2.1] (Section 2.1 Basic requirements) provides one definition of robustness as "A structure shall be designed and executed in such a way that it will not be damaged by events such as explosion, impact, and the consequences of human errors, to an extent disproportionate to the original cause" and that definition will serve well enough.

In previous times, when buildings and their components were sized by rule of thumb, a measure of robustness tended to be built in, albeit not always successfully. In modern times our building methods have become more complex so robustness has ceased to be an attribute taken for granted and a need has arisen to consider it more explicitly. That need is still evolving as the construction industry develops ever more sophisticated and structurally efficient products, and as pressures intensify to build ever more quickly. The evolution of the specialist designer, introducing fragmentation of the design process, is another issue.

In day-to-day work, structural engineers design on the basis of component strength and stiffness against a set of applied loads. That process masks some underlying principles of sound building and tends to obscure the need to look at the whole. Attributes that might be additionally considered are those such as global stability, global stiffness, or insensitivity to settlement, moisture or thermal movements and minor alterations or insensitivity to the inevitable errors that accompany routine construction – all these attributes might be grouped under the quality of robustness or stability. Building components degrade throughout their life at varying rates yet it is not accepted that such decay should lead to any fundamental change in state, at least not without warning. Rather, an underlying safety principle of all sound building is that as far as possible, impending collapse should be signalled by gross deformation or excessive cracking indicating the onset of failure. That same principle applies under overload i.e. that as far as possible structural failure under overload should be ductile rather than brittle. This latter principle applies equally in fire; a hallmark of fire resistance should be stable deformation before collapse.

In the 1955 Report on Structural Safety[2.2] the Institution committee chaired by Sir Alfred Pugsley offered the timeless observations:
– That the structure shall retain throughout its life, the characteristics essential for fulfilling adequately the purpose for which it was constructed, without abnormal maintenance cost.
– That the structure shall retain throughout its life an appearance not disquieting to the user and general public and shall neither have nor develop characteristics leading to concern as to structural safety.
– That the structure shall be so designed that adequate warning of danger is given by visible signs and that none of these signs shall be evident under design working load.

2.2 Hybrid structures

Most structures are a hybrid of different structural materials and forms. For example, timber trusses sitting on masonry, precast concrete units sitting on steel or concrete, timber buildings supported on concrete podiums, or a lightweight steel-framed storey constructed on top of a traditional masonry building. Consequently it is important to emphasise the role of the design team in assuring not only the robustness of the individual material components but also of the structure as a whole (Reference 2.3 exemplifies the dangers).

In all cases it is imperative that the designers co-operate, and co-ordinate their designs, such that the interfaces between materials are robust. This requires procedural co-ordination (e.g. timing of appointments, communication of assumptions and clarity of responsibilities at the interface) and physical co-ordination (e.g. relating to fit, structural continuity or physical access). Section 2.13 talks about the need for one designer to have overall responsibility for stability and robustness and who can take an overview; this is also emphasised in Reference 2.4. Chapter 5 of this *Guide* sets out the design fundamentals.

Box 2.1	The robustness of a structure has to be looked at as a whole
A key presumption is that for any one building, there should be one engineer in overall charge of both stability and robustness and not least when multiple structural disciplines are involved as in hybrid structures. The Task Group have strongly endorsed this principle.	

Thus when considering tying rules, as described in later chapters, the routes of vertical ties to foundation or other appropriate levels should be continuous through the various parts of the structure i.e. across the boundaries of the hybrid structure. If partial collapse is considered, the capability of the structure below the collapse (perhaps designed by others) to support the debris from above must be confirmed.

2.3   Structural form

Achieving these desirable states should start with the concept of structural form. Every structural form should carry gravity and horizontal loads safely to ground and the route those loads take (the load path) should be clearly defined. Whilst often self evident, there are many instances of confusion in older structures and in conversions and even within modern structures where diverse routes are envisaged via floor plates (diaphragms) and bracings or shear walls. A fundamental feature of robustness is that such load paths should be defined.

Once the load path has been defined it should be used to ensure the continuity of a horizontal load resistant path. Although seismic design is not a UK requirement, earthquakes apply horizontal loads to structures and observation of building performance provides empirical feedback over structural form. Time and again the vulnerability of low rise buildings with inadequate lateral resistance in the lowest storey has been demonstrated (so called soft storey collapses). Such failures can illustrate the consequence of wall removal in building conversions which often renders the structure less stiff and less stable than it would have been originally.

Particular care needs to be exercised in hybrid structures (as described in Section 2.2). The assumptions of the various teams on 'what is propped off what', 'what the load paths are' and 'what the stiffness demands are (compatible as between different materials)', all need to be defined very early on and conveyed to the parties involved. There are dangers during the construction phase if the construction sequencing is at odds with overall stability and robustness demands.

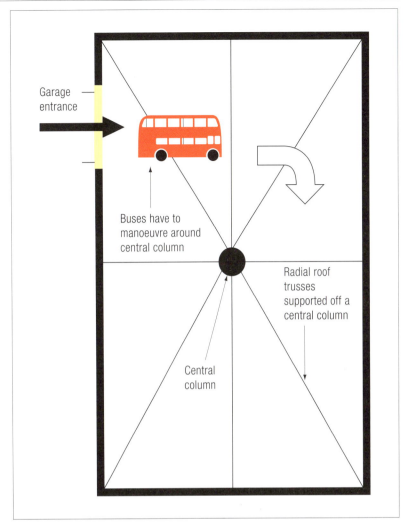

Figure 2.1   Bus garage with roof supported off a central column

Box 2.2	Robust structures are generally stiff structures

The concept of load path aids comprehension about lateral stiffness, the longer the load path, and the higher the stress in the members, the more flexible the structure. Stiffness requires that horizontal loads are taken by the shortest route possible down to ground (see Reference 2.5)

It is not good practice to start off with a structural form which has obvious weaknesses to an implicit hazard associated with the building's function. Figure 2.1 shows a design for a bus depot. The roof was large spanning, but all trusses were propped off a single column located in the middle of the depot. Clearly all the buses had to manoeuvre past this single column and the wisdom of adopting that structural form without careful consideration of its protection can be questioned.

It is generally accepted that the provision of redundancy or some alternative load paths is a good investment to help ensure robustness. In severe overload (as in structures subject to earthquake loading or explosion) the more alternative structural support systems that exist the better. After the Ramsgate collapse[2.6] it became mandatory to provide a catch ledge or chains to support link bridges in case the main support failed. As a generality, caution is required whenever overall integrity is reliant on a single joint.

It is also generally accepted that the more ductile a structural form is, the more robust it is. Ductility desensitises the structure overall or its individual elements to damage from the uncertainties inherent within the loading and service conditions and it desensitises them to variations from predicted stress levels. Ductility is relevant at component level (see Section 2.5) and at structural system level (see Section 2.9). It is advantageous under static loading but is also key to structural response under dynamic loading, in which case it is linked with energy absorption.

2.4   Horizontal loads

The concept of applying horizontal loads underpins many strategies for evaluating the overall robustness of a structural form. Whilst it seems obvious that there should be clear routes for horizontal loads down to the foundations, in many structures it is not immediately obvious what the horizontal loads are in the first place. The Ronan Point failure[2.7] was initiated by a domestic gas explosion applying a horizontal load and there have been plenty more of these since to remind us of the ever present threat. For most structures, it will generally be the case that strong winds provide the major horizontal destabilising force. But horizontal forces also arise due to self weight, perhaps due to side sway from eccentricity of vertical load or perhaps due to inevitable column plumb tolerance. Thus modern codes apply notional lateral loads related to vertical loads. These notional loads can prove more critical than wind loading when applied along the length of long narrow buildings.

Some horizontal loads only apply during construction (see examples in Box 2.3 and Box 2.4).

Box 2.3	Robustness requires that all structures have a resistance to lateral loading and if none is readily defined, a notional percentage of the vertical load is a good starting point (however that percentage is derived).

When a structure is entirely erected within another building, it is not obvious whether there is any horizontal loading to apply at all. But if the structure has no lateral resistance, it will sway and there have been cases of high bay warehouse racking toppling sideways. Imperfections in vertical alignment or partial live loading on the span will all cause structural sway. Such partial loading can exist during construction when floor concrete is being poured from one side. In the permanent case that concrete may form a diaphragm, but during construction it may itself be the destabilising load.

The figure shows a mezzanine floor that was added within a factory and used to store heavy cable drums. Clearly a slight lean of the columns would have generated significant lateral force.

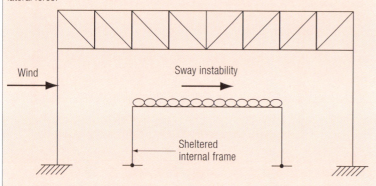

A sheltered internal structure with heavy floor loading yet no defined explicit horizontal load to assure stability

Box 2.4	Robustness requires that there is always a load resisting system for horizontal loading during all stages of construction and this may have to be provided via temporary works.

During construction, many elements that are eventually sheltered within the completed building are temporarily exposed to wind and there have been several instances of walls blowing over before building completion. There have also been instances of tall bridge girders toppling over in the period before the deck has been added.

2.5   Component design

Much of modern structural engineering relies on using components that are rated at maximum efficiency with such efficiency gained by all elements contributing structurally in some fashion. A good example is the steel pitched portal frame with haunched eaves where the lower eaves compression flange is restrained out of plane by proprietary braces back to lightweight cold rolled purlins, which are themselves restrained by the roof covering and in-plane sag rods. Such systems are economical and wide spread. However, care must be taken in use not to remove some of these apparently minor components in case that removal has more widespread consequences, potentially setting off a chain reaction of collapse.

In the absence of buckling, beams are robust under vertical load if they have strong connections and if they exhibit ductility; this is assured by codified rules (such as controlling reinforcement percentages in concrete members or the elimination of local flange buckling in steel). Where failure is by buckling, robustness requires that lateral stability is not dependent on a few flimsy bracings which might raise the danger of gross failure linked to loss of minor members. The cost of bracing is often trivial in comparison with the structure as a whole, so it is not good practice to skimp on it. Likewise, it is not good practice to skimp on beam end connection capacity; whatever the design forces, there should be some correlation between beam capacity and the connections that support it (including the need for connections to act as ties).

Like robustness, ductility is a sound attribute to have and achieving it is partly a matter of design and partly a matter of detailing. Without ductility, structures would be vulnerable to brittle failure and we could not rely on procedures such as slab yield line analysis or plastic design. At a simple level, ductility allows constant shear to be carried as in a real hinge, and a sound objective of steelwork detailing is to allow shear connections to deform yet still carry normal loading even under working conditions. At a more advanced level, ductility allows moment to be carried at constant magnitude when deformation under plastic hinge conditions takes place. The objective of detailing rules is to assure that the plastic hinge has sufficient capacity to permit load redistribution while undergoing further deformation. This can be quantified through the shape of the moment-curvature (M-φ) curve, specifically the part of the curve after maximum moment capacity is attained.

The area under the M-φ curve can also be considered as a measure of the component's ability to absorb energy. Through this concept, we can assess structures with respect to shock dynamic loading, such as occurs in gas explosions, blast, vehicle impact and earthquake action, since all those incidents release a finite amount of energy. Consequently, quantifying and mobilising energy absorption capability is a key ingredient in robustness strategies against accidental loading.

The practice of adopting minimum sizes and minimum slenderness ratios assures a certain amount of construction robustness and many material codes give some guidance over minimum member sizes, minimum amounts of reinforcement, minimum connection capacities, minimum bearing lengths and so on.

Self evidently a robust structure can only be achieved if the quality of detailing and construction is commensurate with the design intent.

Box 2.5	The ability to absorb energy is a key quality of robustness

Looking at dynamic effects in terms of applied force suggests the forces involved can be very high with corresponding high stress, which is alarming yet misleading; a high force exists but it will be of very short duration. A good illustrative example is to investigate the behaviour of a standard crash barrier post under vehicle impact. The impacting vehicle has a certain amount of energy ($0.5mv^2$). When the vehicle hits the post, it bends forming a plastic hinge at its base. The bulk of the vehicle's energy is absorbed by $M_p\theta$. The amount of deflection at the top of the post is then height $\times \theta$, and the designer has some control over deflection magnitude by increasing or decreasing the value of M_p selected: the more deformation acceptable, the lower M_p can be but it cannot be too low otherwise the post would deform through 90° and fail to stop the vehicle.

Moreover, this post design is only valid if the post is capable of significant plastic deformation. To sustain that deformation, the base connection capability is crucial. It has to be able to resist M_p and to sustain M_p throughout the rotation θ. This form of ductility is one characterised by energy absorption, and robustness requires that at least some of the structural system is capable of acting in this way.

Other parts of the system may only need to exhibit rotational ability without absorbing energy (as for example in a door hinge), such joints only need carry shear whilst deforming significantly without fracturing. It will be found that the concepts of ductility, energy absorption and rotational capacity and how these are achieved by design and detailing underlie many advanced studies of structural robustness. The example of the crash barrier post is good analogy for the demands made on structural systems as a whole.

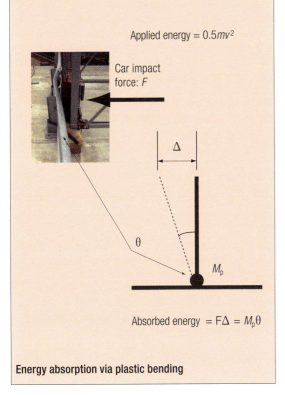

Applied energy = $0.5mv^2$

Car impact force: F

Δ

θ

M_p

Absorbed energy $= F\Delta = M_p\theta$

Energy absorption via plastic bending

2.6 Redundancy

When failure of any one loadbearing member leads to the collapse of the entire structural system, as is the case in pin-jointed trusses, the structure has no redundancy. Statically determinate structures can revert to a mechanism if just one part fails. In this case, the structure can be thought of as a chain under tension – loss of one link implies loss of the ability to transfer load from one end to the other. This is known as a series or weakest-link system. Clearly, load transfer cannot take place in series systems when failure has occurred and this is undesirable from a robustness perspective.

On the other hand, in a parallel system, members are interconnected in such a way that load is shared between them, and failure of one member leads to load redistribution to other members. Such a system could have a degree of redundancy – consider a vessel anchored through four mooring lines and the consequences of one (or more) failing. The degree of redundancy will depend on whether the re-distributed load can be taken up by the remaining members. In turn, this will be a function of whether the failed element has retained an ability to carry some load. Using the moored vessel analogy, the failed line could snap (brittle failure – only three lines are now carrying the load) or could yield (ductile failure – the failed line still carries some load but anything in excess of yield is distributed to the remaining three).

In parallel or redundant systems, further distinctions can be made depending on whether redundant members pick up loads even when the structure experiences a low level of loading (active redundancy) or whether they participate only after a certain degree of degradation/damage has occurred (passive or fail-safe redundancy); back up tie cables can be used to secure objects that might fall due to fatigue or corrosion etc. A second important factor is the so-called common cause failure, in other words whether members that are expected to share loads are susceptible to common underlying factors that may lead to their failure at the same time or under the same conditions (e.g. all mooring lines suffer from the same manufacturing defect).

One strategy of Building Regulations and codified methods of imparting robustness is to rely on a level of structural continuity and redundancy. When this is provided there may be sufficient capacity in any undamaged structure to carry loads redistributed from the damaged elements often via alternative load paths. In normal structural engineering terms, redundancy means the structure has more load paths than it strictly needs. Inherently this is a 'good thing' since damage or deficiency in any one part (joint or component) does not instantly spell failure in the accepted sense.

In any discussions on robustness there is a second level of redundancy or spare capacity that can be exploited: the available margin existing between what the elements are capable of carrying and what the demand actually is. Margins exist as the difference between real material strength and specified strength; and margins exist because the loads on the structure during the initiating event will often be less than those used for design purposes. Codes offer reduced safety factors to allow for these facts.

Box 2.6	Robust structures are insensitive to the precision of design assumptions

When designing for wind as the dominant load, care must be taken that the structure is not too sensitive. For example, it is customary for a reduced wind load to be taken during construction, but at a low speed of say 10m/sec, a credible marginal increase of 1m/sec will raise the forces by $(11/10)^2 = 21\%$. Coping with such potential variance requires robustness, not more precision in the calculations.

Box 2.7	Robust structures are insensitive to construction accuracy

The figure shows a short cantilever. Nominally the applied load on the connection bolts is a shear load P and a BM of P x 20mm. But a credible 20mm error in construction tolerance will result in a BM of P x 40mm which is a doubling of the presumed moment. Robustness requires that consideration be given to credible construction error. Clearly if the cantilever was 1m long and the tolerance remained at 20mm, the effects would be much less.

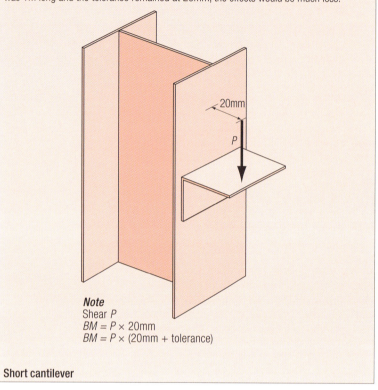

Note
Shear P
$BM = P \times 20mm$
$BM = P \times (20mm + \text{tolerance})$

Short cantilever

Box 2.8	Designers need to consider the sensitivity of their calculated results to variations in their design assumptions

The figure shows a thin concrete slab that was designed as a post tensioned slab with the tendon offset by 10mm to achieve the requisite bending capacity. However, the designer had not considered that a perfectly credible tendon positioning error of 10mm would totally undermine the assumptions made and fatally weaken the slab: the design was not robust.

As with the figure in Box 2.7, all credible variations of small dimensions should be considered.

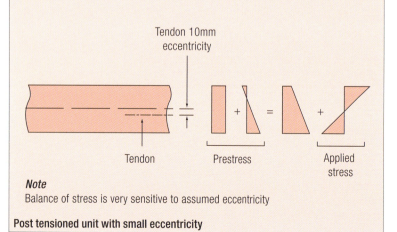

Note
Balance of stress is very sensitive to assumed eccentricity

Post tensioned unit with small eccentricity

2.7   Insensitivity/scale

The concept of insensitivity is an attribute of robustness and this can be seen in many ways. The description of the variation in wind load in the example in Box 2.6 is just one example of a design being too mathematical and too sensitive to assumptions that could prove inaccurate: as a generality, all loading magnitude is uncertain and engineers have to guard against that.

Design care is required when the absolute errors of tolerance or building movement can have a disproportionate effect. Refined calculations are inappropriate if they neglect this fact. The examples in Box 2.7 and Box 2.8 illustrate the point. The converse to the boxed examples might occur in design when axial loads are considered concentric, it is always sensible to assume some accidental eccentricity.

As a principle, designs should account for credible variations in design assumptions. This applies, say, in reinforced concrete design where minor bar positional errors are a fact of life, or in steelwork where joint shimming might be required. There have been cases of sudden failure of flat slabs in shear around column heads where bars have been mis-positioned or trampled down during concreting. It is well known that a 5mm loss of cover will have a significant deleterious effect on the durability of reinforced concrete. Good design requires that insensitivity is achieved by good detailing, by recognising the difficulties of construction and by concentrating on such matters at the expense of structural efficiency expressed purely in numerical terms of strength.

Classic structural failures that might be cited as an example of lack of robustness are the failure of Rock Ferry School Sports Hall[2.3] in 1976 and the collapse of Camden School assembly hall roof[2.8]. Both failures illustrate sensitivity to details. See Box 2.9 and Box 2.10.

2.8   Uncertainty

Structural engineering should never become a totally mathematical exercise; the real strength of structures is a function of their theoretical design, the quality of detailing and the quality of construction. There are uncertainties in each of those stages. References 2.9 and 2.10 discuss some of the fundamental uncertainties along with the need for ductility to overcome them. The failure in Box 2.11 illustrates one aspect of loading uncertainty.

| Box 2.9 | Robustness is a quality influenced by detailing quality |

Rock Ferry School Sports Hall and Camden School Hall

The gymnasium of Rock Ferry School had loadbearing masonry walls and the roof was made of timber trusses seated on the wall. Although the timber trusses were interconnected, there was in effect no bracing out of plane and no load path for longitudinal horizontal (wind) loads back to ground. In the event, the trusses toppled sideways and the whole roof collapsed. The failure of the hall roof at Camden school in 1973 was similarly catastrophic but was precipitated by the trivial bearing width provided for the roof beams and by corrosion of the reinforcement that was supposed to keep the beams on the support.

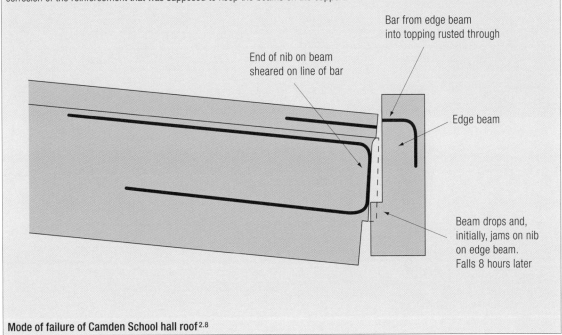

Mode of failure of Camden School hall roof[2.8]

| Box 2.10 | For structures to be robust, they should be insensitive to common building movements |

Robustness requires the question being asked about what might happen if supports move, say, due to settlement or differential thermal movement over time. The figure shows a floor in a multi-storey building consisting of precast flooring units spanning onto edge beams. The calculations showed the vertical capacity of the units to be perfectly adequate but the designers had not considered the possibility of the side units bowing out of plane, either increasing torsion on the edge supports or worst still, allowing them to move far enough apart to let the units drop through. *In situ* ties were required at intervals simply to stop this happening. The needs of safe erection also dictate that designers consider any out of balance loadings that might exist during flooring installation.

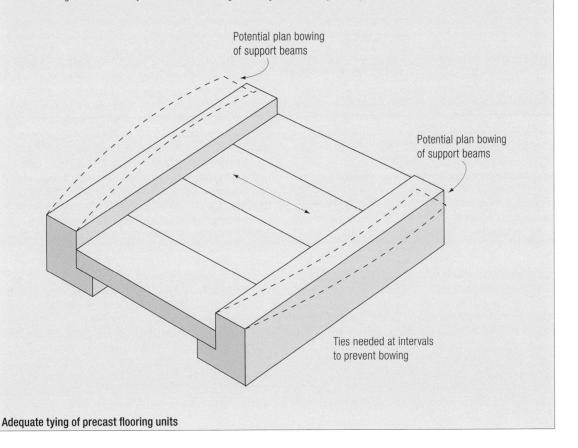

Adequate tying of precast flooring units

Box 2.11	Robustness concepts are important for any assembly of building components

The figure shows a suspended grillage ceiling which supported an air conditioning plant and services to a magnitude of about 1kN/m². The application of loading was indeterminate since the ducting was rigid, spanning over some supports and concentrating loads on others. The whole grid was held up by proprietary anchors drilled into a concrete soffit. In service, one of the anchors pulled out, subsequently overloading the adjacent ones which then also failed and a cascade was set up whereby the whole grillage came down.

This is an example of both progressive and disproportionate collapse. Robustness demands that when possibilities for failures of this kind exist, there has to be high confidence that an initiating event cannot occur. In this case, site testing of anchor installation would have been prudent. As the distribution of loading was fundamentally indeterminate, a good safety margin was required on notional anchor pullout capacity to cover that uncertainty. Designers should always be conscious of the accuracy with which they can predict loading and structural capacity.

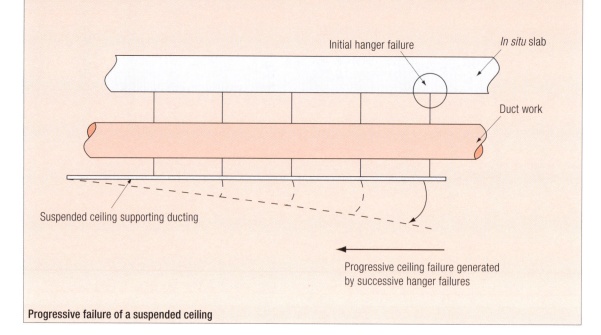

Initial hanger failure

In situ slab

Duct work

Suspended ceiling supporting ducting

Progressive ceiling failure generated by successive hanger failures

Progressive failure of a suspended ceiling

2.9   Failure modes and ductility

As mentioned in Section 2.5, ductility is a sound quality for a structure to have if it is to be robust. Most codes assure a level of useful ductility at component level by imposition of detailing rules. Structural ductility allows parts to deform yet still carry load – it allows overloaded parts of the structure to yield and redistribute stress.

The Building Regulations and codified methods of imparting robustness implicitly rely on a level of ductility within the structure. Catenary action, which is a fundamental assumption of some survival strategies, relies on the ability of connections and joints to deform without fracture and, for example, reinforcing bars to elongate without fracture. In routine design, such ductility is not explicitly evaluated, instead achievement has to be assured by good detailing which means using proven detailing techniques normally as recommended by relevant trade organisations.

It is implicit within the concepts of ductile response that structures may distort significantly under accident conditions; this is tolerable provided that the structure stays intact and does not collapse completely, although limited area collapses are permissible. The presumption of significant distortion is only justified if the structure has been appropriately detailed.

Ductility at structural system level is implicitly linked with energy absorption capability. For example, large frames under severe dynamic loading can be designed so that certain failure modes, which involve a particular sequence and location of plastic hinge formation, can occur in preference to others. In this way, the energy absorption capability of the structure can be enhanced, thus allowing more severe dynamic events to be contained. To allow the formation of these preferred failure modes, connection ductility should be properly assessed and catered for. It is also important in this case to avoid any inadvertent switch of failure modes by providing additional but unwanted strength to elements that have been assumed to participate in the preferred failure modes. This strategy is used in earthquake resistant design, where a weak beam-strong column design has to be adhered to so that sway failure modes with considerable energy absorption capability can develop in preference to soft-storey modes which dissipate energy in a local part only.

In general, for routine building work it is not necessary to make any explicit calculations on ductility. Explicit calculations are used in seismic design and in blast resistant design, for which specific detailing methods that assure high ductility in the context of system failure can be found in textbooks and codes. Although there is no need to consider these in routine design, an understanding of the background and an intuitive application will certainly help in more routine structural work.

2.10 Fire loading and robustness

Fire is an important accidental load case and thus it is essential that due consideration is given to the robustness of a structure in the event of a fire. Some of the objectives are:
– to design a structure such that a fire does not cause a disproportionate failure
– to maintain stability for a period sufficient to allow means of escape to occur
– to maintain stability for a sufficient period to enable fire fighting and search and rescue.

For the majority of small scale buildings, the prescriptive requirements defined in Approved Document B[2.11] and BS 9999[2.12] will be sufficient and no special design features will be necessary. However, the responsible structural designer should maintain an awareness to confirm that there is nothing unusual that would require additional attention. Those responsible for the fire safety design strategy should be able to provide this advice.

The need for a robustness assessment will be greater as the horizontal and vertical dimensions of the building increase and thus there will be an increasing need to co-ordinate the fire safety and the normal structural design cases.

Where a need is identified to review the robustness in more detail, the following considerations will impact on the assessment:
– An appropriate form of the structure and level of redundancy will reduce the chances of a disproportionate event affecting life safety.
– The details and the ductility of the structure and the type of connections will impact on the ability of a building to resist collapse during both the heating and the cooling phases.
– The presence of alternative load paths (e.g. catenary action under large deflection) has a significant impact on the fire load case and the ability of the structure to continue to support load at elevated temperatures.
– In many instances, the larger deflections that result during a fire are acceptable as long as collapse is not initiated.
– Consideration should be given to real fire performance rather than the approach dictated by tests and the use of the standard fire curve.

The long term performance of a structure is equally important as a fire can occur many years after construction so a fire safety solution needs to be sustainable and maintainable. The communication of this information is required by Approved Document B[2.11] (as specified by Regulation 16B of the Building Regulations).

For more detailed consideration see References 2.13 and 2.14.

2.11 Progressive and disproportionate collapse

The Ronan Point failure was the classic example of progressive collapse; that is the failure of one member which set off a chain reaction of other collapses such that the totality of damage was quite disproportionate to the initiating event. No engineer can prevent total collapse if the event is big enough, but a robust structure should assure that the extent of damage is not disproportionate to the initiating event. The regulations that will be discussed in the following chapters aim to provide rules that will assure the containment of any damage to prescribed amounts.

2.12 Summary

Key points relating to robustness are:

(1) Robustness is a 'good thing' to protect structures against the unforeseen.

(2) Robustness is not purely a mathematical quality that can be measured in terms of strength.

(3) Robustness of structural systems starts by having a structural form appropriate to building function. Attributes such as redundancy, alternative load paths and clarity of load path are all good to have and in some cases essential.

(4) Robustness is achieved by making the structure strong yet ductile with ductility having an importance comparable to strength. Joints are crucial to robust performance. Where reliance is placed on a single joint, special care is required.

(5) Robustness requires that all structures have adequate load paths down to the foundations for vertical and horizontal loads (in each orthogonal direction). In some cases, horizontal loads should be notional, related to dead and imposed loads.

(6) A measure of a structural system's robustness is its insensitivity to any change in state consequent upon a credible variation of any design assumptions. Structural capacity should not be sensitive to variations in assumptions such as load positioning, construction accuracy, tolerances or in-service degradation.

(7) Designers should always be mindful of the hazards that are relevant to building/structural function.

(8) In any structure there should be one engineer who has responsibility for the provision of overall stability and overall robustness.

2.13   Strategy

In all structures, and particularly in hybrid structures, there must be one designer who takes primary responsibility for assuring stability and robustness of the whole structure and who defines and documents the strategy. This principle does not remove the need for other individual designers to take responsibility for the robust design of the various elements that make up the whole. Equally the same point applies during construction; the risks of failure are highest in the partially completed state and one designer should have overall responsibility for assuring robustness/ stability during the construction phase. In reviews for design and construction, the question over how robustness will be tackled ought to be raised.

The strategy for achieving robustness lies partly in the competence of the design team, partly in compliance with good building practice and partly in compliance with regulations and codified rules. There are various documents which define overall strategies (for example BS EN 1990[2.1]) and Figure 2.2 reproduces the flow chart from BS EN 1991-1-7[2.15] (Figure 3.1 from the source). This leads on to detailed strategies already incorporated in British practice as described in later chapters of this *Guide*.

It has to be emphasised here that although much UK regulation stems from the accident at Ronan Point, the initiating event now underpinning the UK design processes is a purely notional value whereas in BS EN 1991-1-7[2.15] the structure may be designed for specific actions as indicated in the figure extract below. Notwithstanding the UK rules, designers should always consider specific risks that might apply to their structures alone. Proximity to heavy moving vehicles is a common occurrence.

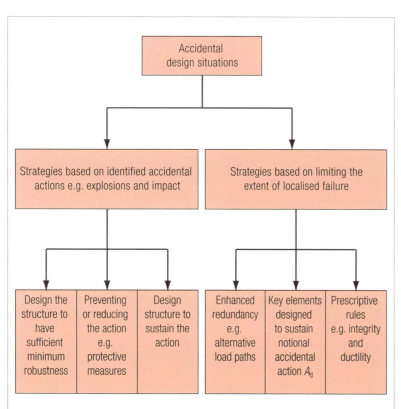

Figure 2.2   Strategies for Accidental Design Situations (Figure 3.1 from BS EN 1991-1-7)

2.14   References

2.1   *BS EN 1990: 2002: Eurocode: Basis of structural design*. London: BSI, 2002

2.2   'Report on structural safety'. *The Structural Engineer*, 33 (5), May 1955, pp141-149

2.3   Menzies, J.B. and Grainger, G.D. *Report on the collapse of the sports hall at Rock Ferry Comprehensive School, Birkenhead. BRE Current Paper CP 69/76*. Garston: BRE, 1976

2.4   Whittle, R. and Taylor, H. *Design of hybrid concrete buildings: a guide to the design of buildings combining in-situ and precast concrete*. Camberley: The Concrete Centre, 2009

2.5   Ji, T. 'Concepts for designing stiffer structures'. *The Structural Engineer*, 81(21), 4 November 2003, pp36-42

2.6   Chapman, J.C. 'Collapse of the Ramsgate Walkway'. *The Structural Engineer*, 76(1), 7 January 1998, pp1-10; Discussion, 78(4), 15 February 2000, pp22-29

2.7   Ellis, B.R. and Currie, D.M. 'Gas explosions in buildings in the UK: regulation and risk', *The Structural Engineer*, 76(19), 6 October 1998, pp373-380

2.8   Department of Education and Science. *Report on the collapse of the roof of the assembly hall of the Camden School for Girls*. London: HMSO, 1973

2.9   Beeby, A.W. 'Safety of structures and a new approach to robustness'. *The Structural Engineer*, 77(4), 16 February 1999, pp16-21

2.10   Heyman, J. 'Hambly's paradox: why design calculations do not reflect real behaviour'. *ICE Proceedings, Civil Engineering*, 114(4), November 1996, pp161-166

2.11   Department for Communities and Local Government. *The Building Regulations 2000. Approved Document B: Fire Safety. Vol 1: Dwellinghouses; Vol 2: Buildings other than dwellinghouses*. 2006 ed. London: NBS, 2007

2.12   *BS 9999: 2008: Code of practice for fire safety in the design, management and use of buildings*. London: BSI, 2008

2.13   Institution of Structural Engineers. *Introduction to the fire safety engineering of structures*. London: Institution of Structural Engineers, 2003

2.14   Institution of Structural Engineers. *Guide to the advanced fire safety engineering of structures*. London: Institution of Structural Engineers, 2007

2.15   *BS EN 1991-1-7: 2006: Eurocode 1: Actions on structures – Part 1-7: General actions – Accidental actions*. London: BSI, 2006 and *NA to BS EN 1991-1-7: 2006: UK National Annex to Eurocode1 - Actions on structures – Part 1-7: Accidental actions*. London: BSI, 2008

3 Legal and other obligations

3.1 Introduction

Structural engineers are legally obligated to design structures that are safe to construct, use, maintain and de-commission/demolish. Part of this obligation is to ensure stability/solidity/robustness and to guard against disproportionate collapse. This chapter outlines the background and gives guidance as to how the designer's obligations may be discharged.

Although obligations will vary around the world, the underlying requirements and principles are likely to be similar to the UK requirements regardless of the structure's location hence the following commentary only references UK practice. The generic categories of obligation in the UK are shown in Table 3.1. These are not intended to be ranked although statutory provisions always take precedence.

Table 3.1   Categories of obligation

	Category	Comment
A	Statute	Legislative requirements
B	Contract	Between employer/employee and between employer/client
C	Common law	Based on case law creating a duty of care
D	Code of Conduct	Institution of Structural Engineers Code of Conduct requires members to have regard to the public's safety; designers belonging to other Institutions are likely to have similar obligations

3.2 Category A: Statute

Category A legislative requirements are shown in Table 3.2.

The Building Regulations in the UK are promulgated under three different jurisdictions: England & Wales, Scotland, and Northern Ireland. However, the requirement concerning robustness and disproportionate collapse is essentially the same in all three, although the limits of application and some definitions vary. The regulations in England and Wales[3.4] are currently the responsibility of the Department of Communities and Local Government. The same department also publishes Approved Document A[3.5] (AD-A) which is the key reference.

In the following text, reference is generally limited to the legislation and guidance for England and Wales. Regardless of the jurisdiction, the aim of the Building Regulations is the same. Regulation 8[3.4] offers helpful clarification. Headed "Limitation on requirements", it states "Part A [inter alia] shall not require anything to be done except for the purpose of securing **reasonable standards of health and safety** (our emphasis) for persons in or about buildings (and any others who may be affected by buildings or matters connected with buildings)".

As can be seen from Table 3.2 the Workplace Regulations[3.13] relate to stability and solidity rather than robustness. However the latter is a useful general phrase to use and is applicable to all structure types.

Table 3.2   Source of statutory obligation

Statute	Comment
Building Act 1984[3.1] Building (Scotland) Act 2003[3.2] Building Regulations (Northern Ireland) Order 1979 (as amended 1990 and 2009)[3.3]	Enabling legislation allowing for more detailed regulations to be made
Building Regulations 2000 (as amended 2004)[3.4]	Requires the design of relevant buildings to satisfy Part A 'Structure'. Official guidance is given in Approved Document A[3.5], and item A3 in particular
Building (Scotland) Regulations 2004[3.6]	Supported by Technical Handbook[3.7] Section 1
Building Regulations (Northern Ireland) 2000[3.8]	Supported by Technical Booklet D[3.9]
Health and Safety at Work etc. Act 1974[3.10] Health and Safety at Work Order (Northern Ireland) 1978[3.11]	Requirement under s3 to have regard for those affected by the designer's work (i.e. their undertaking). Typically, 'those affected' will be contractors, users, and the public As above (under Article 5)
Construction (Design and Management) Regulations 2007 (CDM) [3.12]	Requirement to 'eliminate hazards and reduce risks from any remaining hazards' and to provide relevant information (Regulation 11) Applies to all foreseeable risks
Workplace (Health, Safety and Welfare) Regulations 1992 (as amended) [3.13]	Requirement for workplaces to have 'stability and solidity' (Regulation 4A)

Notes
a These regulations are amended or revoked from time to time, hence it is important to ensure current versions are being used.
b Safety regulations in Northern Ireland are identical in the detail to those elsewhere in the UK.

Building Regulation A.3[3.4] requires that: "the building shall be constructed so that in the event of an accident the building will not suffer collapse to an extent disproportionate to the cause". Approved Document A[3.5] provides one means of meeting compliance with this key regulation. The specified measures (described elsewhere in this *Guide*) are stipulated to ensure adequate robustness and safeguards against disproportionate collapse of buildings arising from the effects of accidental events; malicious action is not covered by the Building Regulations.

A 'building' is defined in the Building Act[3.1] (Section 121) as (part extract) "Any permanent or temporary building, and, unless the context otherwise requires, it includes any other structure or erection of whatever kind or nature (whether permanent or temporary)".

Approved Document A[3.5] allows any method of structural analysis and design to be adopted providing it meets the relevant requirement of the regulations[3.4]. However, most submissions made under the Building Regulations utilise the scheduled list of approved codes of practice as the means of demonstrating compliance. However, not all building types will fit with the assumptions inherent in these codes, and compliance with British Standards and other codes does not confer immunity from legal obligations.

Approved Document A[3.5] (Table 11) categorises buildings for the purpose of designing for robustness and against disproportionate collapse. These classes (which do not cover all categories of building e.g. stations and surgeries) are described in Chapter 4. Classes 1, 2A, 2B have prescriptive solutions presented, whereas the acceptable solutions for Class 3 are to be determined by the designer from an assessment of risk. Although important, Class 3 buildings are not considered further in this *Guide* (though guidance is available in Reference 3.14).

In general, the Building Regulations[3.4] only require the effects of accidental events to be applied to the completed building. However, if a material alteration is undertaken on an existing building then the inherent level of robustness is required to be maintained during the construction phase of this work as well. Notwithstanding this formality, Statute (the Health and Safety at Work etc. Act[3.10]) requires consideration of **all** foreseeable risk (i.e. accidental and malicious) at **all** stages of the building's life i.e. construction, use, alteration and repair and decommissioning. Such consideration would normally be discharged by competent designers identifying and eliminating hazards so far as is reasonably practicable, and then reducing residual risks from any remaining hazards to which the structure is reasonably exposed. Hence structural engineers should prudently consider the construction phase since that is the stage when accidental damage or instability behaviour is most likely to occur. Risk reduction can be achieved by:
– following good practice which is authoritative and accepted by the Health and Safety Executive (HSE) as such, or
– working from first principles (and taking account of the general principles of robustness discussed throughout this *Guide*).

Risks in the context of health and safety legislation include not only the consequence of physical actions

e.g. vehicular impact, but could also for example include the consequences of inadequate analysis or design, or risk arising from a lack of competency of the persons used in the design/construction process. Such risks must be managed. BS EN 1990[3.15] (and reinforced elsewhere in the Eurocodes) also emphasises this broad approach to risk.

The CDM Regulations[3.12] reinforce the requirement to consider robustness during construction and de-commissioning as well as the operational in-use phase covered by the Building Regulations[3.4]. It is important that the designer's intent and assumptions are conveyed to others as noted in the penultimate paragraph of this Section below.

The Workplace (Health, Safety and Welfare) Regulations[3.13], as amended, require workplaces in buildings to have 'stability and solidity'. The regulations offer no formal guidance. However, it would be reasonable to conclude that, for design purposes, this would be achieved by:
– compliance with the Building Regulations
– any other relevant risks being accommodated
– the structure being maintained to a reasonable structural standard
– the use of the workplace being as originally envisaged by the designer, and
– having any significant change to the structure or its use being reviewed and assessed against robustness criteria.

The requirements of the Health and Safety at Work Act[3.10] and the CDM Regulations[3.12] apply to buildings and all other types of structure. Compliance is to be achieved 'so far as is reasonably practicable' (SFARP) in respect of all foreseeable risks. Guidance is available in Reference 3.16.

A summary of the scope of the various statutory provisions is given in Table 3.3.

The design process is required to include the provision of adequate information to others to ensure robustness and to guard against disproportionate collapse during the construction or subsequently in the life of the structure. This information might involve:
– a description of the structural system, load paths and stability system in the permanent state in sufficient detail that constructors and those in the future who may be engaged to amend or de-commission the structure can appreciate issues of temporary stability
– requirements of temporary bracing or suggested construction sequences and the like during the construction process. From these the contractor may develop a safe system of work
– design assumptions regarding the use and maintenance of the structure
– adequately detailed drawings.

This data would normally be contained in the pre-construction data or in the Health and Safety File if the project is notifiable under the Construction (Design and Management) Regulations 2007[3.12] (i.e. over 30 days or involving more than 500 person days on site) or in other documents if less than these limits. Robustness during the construction phase is particularly important. For this to be achieved, the contractor must be informed of this key information.

Table 3.3   Scope of legislation

Legislation	Applicable to robustness during or at:								Type of structure	Included risks
	Construction	Completion	Use	Work associated with change of use	Maintenance	Structural repair	Structural alteration	De-commissioning		
Building Regulations[3.4]		✓	✓ a	✓ b			✓ c		Buildings	All foreseeable and unforeseeable accident risks (excludes malicious actions)
Health and Safety at Work Act[3.10]	✓	✓	✓	✓	✓	✓	✓	✓	All structures	All foreseeable risks
CDM Regulations[3.12]	✓	✓	✓	✓	✓	✓	✓	✓	All structures	All foreseeable risks
Workplace Regulations (Regulation 4A)[3.13]			✓	✓ d	✓ d	✓ d	✓ d		Buildings that are workplaces	All foreseeable risks

Notes

a   Only in cases where the structure is judged as being dangerous do the provisions of the Building Act 1984[3.1] apply and allow it to be made safe.

b   Only if change is material and relevant, as defined in the Building Regulations[3.4].

c   Only if the alterations are material as defined in the Building Regulations[3.4].

d   Only if workplace remains in use.

3.3   Category B: Contract

Contracts between the engaging party (usually the client for the project or the contractor if a design and build procurement route) and the designer will require – explicitly or implicitly – that designs should satisfy statutory provisions or some more stringent requirement e.g. measures to resist terrorist action (this should be agreed with the engaging party). The standard of care expected in discharging this obligation should be stated in the contract, but it would normally be to undertake the design using due skill, care and diligence. The designer would be expected to be aware of contemporary good practice and industry advice and concerns. Contract requirements are subordinate to statutory requirements but contractual obligations can impose more onerous requirements than are imposed by statute.

3.4   Categories C and D: Common law and Code of Conduct

The common law duty of care (Category C), and professional obligations (Category D) will normally be satisfied by compliance with Categories A and B.

3.5   References

3.1   *Building Act 1984. Chapter 55.* London: HMSO, 1984

3.2   *Building (Scotland) Act 2003.* [s.l.]: The Stationery Office, 2003

3.3   *Building Regulations (Northern Ireland) Order 1979.* [s.l.]: HMSO, 1979 (SI 1979/1709 (N.I. 16)) as amended by *The Planning and Building Regulations (Amendment) (Northern Ireland) Order 1990.* [s.l.]: The Stationery Office, 1990 (SI 1990/1510 (N.I. 14)) and the *Building Regulations (Amendment) Act (Northern Ireland) 2009. Chapter 4.* [s.l.]: The Stationery Office, 2009

3.4   *The Building Regulations 2000.* London: The Stationery Office, 2000 (SI 2000/2531), as amended by *The*

Building (Amendment) Regulations 2004. [s.l.]: The Stationery Office, 2004 (SI 2004/1465)

3.5   Office of the Deputy Prime Minister. *The Building Regulations 2000. Approved Document A: Structure.* London: NBS, 2004

3.6   *Building (Scotland) Regulations 2004.* Edinburgh: The Stationery Office, 2004 (SSI 2004/406)

3.7   Scottish Building Standards Agency. *The Scottish building standards technical handbook: domestic; non-domestic.* Edinburgh: The Stationery Office, 2010

3.8   *Building Regulations (Northern Ireland) 2000.* [s.l.]: The Stationery Office, 2000 (Statutory Rule 2000/389)

3.9   Department of the Environment for Northern Ireland. *Building Regulations (Northern Ireland) 2009. Technical Booklet D: Structure.* London: HMSO, 2009

3.10   *Health and Safety at Work etc. Act 1974. Chapter 37.* London: HMSO, 1974

3.11   *The Health and Safety at Work (Northern Ireland) Order 1978.* London: HMSO, 1978 (SI 1978/1039 (N.I. 9))

3.12   *The Construction (Design and Management) Regulations.* Norwich: The Stationery Office, 2007 (SI 2007/320)

3.13   *The Workplace (Health, Safety and Welfare) Regulations 1992.* London: The Stationery Office, 1992 (SI 1992/3004), as amended by *The Health and Safety (Miscellaneous Amendments) Regulations 2002.* [s.l.]: The Stationery Office, 2002 (SI 2002/2174)

3.14   Harding, G. and Carpenter, J. 'Disproportionate collapse of Class 3 buildings: the use of risk assessment'. *The Structural Engineer,* 87(15), 4 August 2009, pp29-34

3.15   *BS EN 1990: 2002: Eurocode: Basis of structural design.* London: BSI, 2002

3.16   Institution of Civil Engineers. *A review of, and commentary on, the legal requirement to exercise a duty 'so far as is reasonably practicable' with specific regard to designers in the construction industry.* Available at: http://www.ice.org.uk/knowledge/specialist_community_health_downloads.asp [Accessed: 24 February 2010]

4 Regulations, codes of practice and supporting documents and their interpretation

4.1 Introduction

4.1.1 General

This chapter describes the general framework of UK regulations covering issues of robustness and disproportionate collapse. The requirements for disproportionate collapse have a long history within the Building Regulations[4.1], first being introduced following the partial collapse of the Ronan Point flats in 1968[4.2]. Buildings under five storeys were originally exempt, but the current regulations (as amended in 2004) now apply to all buildings. (In Northern Ireland and Scotland revised regulations, post 2009, bring local regulations very close to those of England and Wales.)

4.1.2 Approved Document A

The regulations are promulgated by the Office of the Deputy Prime Minister (since transferred to Communities and Local Government). The same organisation also publishes Approved Document A[4.3] (AD-A). This opens with the statement:

> "This document … has been approved by the First Secretary of State for the purpose of providing practical guidance with respect to the requirements of … the Building Regulations …"

and continues;

> "approved documents are intended to provide guidance for some of the more common building situations. however, there may well be alternative ways of achieving compliance with the requirements. **Thus there is no obligation to adopt any particular solution contained in an approved document if you prefer to meet the relevant requirements in some other way**" (bold type in the original).

Readers should check that they are referring to the current edition of AD-A (www.planningportal.gov.uk).

4.1.3 Eurocodes

From the above it can be seen that AD-A[4.3] is equivalent to a code of practice. The other code which will have wide application is Eurocode BS EN 1991-1-7 *Actions on structures: General actions – accidental actions*[4.4]; the equivalent guidance is in Annex A *Design for consequences of localised failure in buildings from an unspecified cause*. The UK National Annex to BS EN 1991-1-7 does not amend or supplement the EN in any significant way except to identify Annex A as informative. AD-A and EN 1991-1-7 offer very similar guidance, so this will be explained as from AD-A but with variations in Annex A of BS EN 1991-1-7 prefixed with 'EN' (in the following text).

4.2 Classification of structures

For practical implementation, buildings are classified according to the perceived consequences of failure. Their allocation in AD-A[4.3] according to 'building class' ('consequences class' in EN, 'Risk Group' in Scotland) is based on building type, number of storeys and occupancy and can conveniently be set out in a table, see Table 4.1. (The classification in Scotland and Northern Ireland is very similar).

Class 1
- Houses not exceeding 4 storeys.
- Agricultural buildings.
- Buildings into which people rarely go, provided no part of the building is closer to another building, or area where people do go, than a distance of 1.5 times the building height.

Class 2A
- Single occupancy houses exceeding 4 storeys.
- Hotels not exceeding 4 storeys.
- Flats, apartments and other residential buildings not exceeding 4 storeys.
- Offices not exceeding 4 storeys.
- Industrial buildings not exceeding 3 storeys.
- Retailing premises not exceeding 3 storeys of less than $2000m^2$ ($1000m^2$ in EN) floor area in each storey.
- Single-storey educational buildings.
- All buildings not exceeding 2 storeys to which the public are admitted and which contain floor areas not exceeding $2000m^2$ at each storey.

Class 2B
- Hotels, flats, apartments and other residential buildings greater than 4 storeys but not exceeding 15 storeys.
- Educational buildings greater than 1 storey but not exceeding 15 storeys.
- Retailing premises greater than 3 storeys but not exceeding 15 storeys.
- Hospitals not exceeding 3 storeys.
- Offices greater than 4 storeys but not exceeding 15 storeys.
- All buildings to which the public are admitted which contain floor areas exceeding $2000m^2$ but less than $5000m^2$ at each storey.
- Car parking not exceeding 6 storeys.

Class 3
- All buildings defined above as Class 2A and 2B that exceed the limits on area and/or number of storeys
- All buildings to which members of the public are admitted in significant numbers (category in EN only)
- Grandstands (stadia in EN) accommodating more than 5000 spectators
- Buildings containing hazardous substances and/or processes.

Table 4.1   Building classes based on building type, number of storeys and occupancy

Building type	Building Class			
	1	2A	2B	3
Agricultural	All			
Buildings into which people rarely go	H ←Gap→ Other building or people Gap > 1.5H^a			
Houses	1 – 4	5 (single occupancy)	6+ b	
Hotels, flats and other residential buildings		1 – 4	5 – 15	16+
Offices		1 – 4	5 – 15	16+
Retail		1 – 3 **and** < 2000m² (1000m² in EN) floor area per storey	4 – 15 **and** < 2000m² (no area limit in EN) floor area per storey	16+ **or** > 2000m² floor area per storey
Buildings to which public are admitted		1 – 2 **and** < 2000m² floor area per storey	< 5000m² floor area per storey	> 5000m² floor area per storey (significant numbers in EN)
Educational		1	2 – 15	16+
Car parks			1 – 6	7+
Hospitals			1 – 3	4+
Industrial		1 – 3 **and** not containing hazardous substances/ processes		4 **or** containing hazardous substances/ processes
Grandstands (stadia in EN)				> 5000 spectators

Notes

a H is height of building being considered. For buildings not meeting this requirement, treat as a building from the most appropriate use, often industrial.

b Any building of 6 storeys (even if nominally single occupancy, is better assigned to Class 2B).

c This table re-arranges the listing in Table 11 of AD-A$^{4.3}$

The most notable difference between AD-A$^{4.3}$ and BS EN 1991-1-7$^{4.4}$ is the inclusion of buildings to which the public are admitted in significant numbers in Class 3 in the EN. No guidance is offered on what constitutes significant numbers, but this would be expected to be considered in the risk assessment.

There are also buildings of special purpose which do not come into Class 3 by virtue of their size but still merit Class 3 treatment because of their value, vulnerability or because the consequences of failure would be particularly serious. Examples might include:
– animal research centres
– banks, bonded warehouses
– control centres (e.g. for industrial plants, railways)
– data centres
– embassies and high commissions
– government offices
– religious buildings
– leading museums and galleries.

The same is true for innovative structures, and those designing such structures should recognise that by their nature these are more vulnerable to unknown effects. Innovative structures need to be sufficiently robust to tolerate behaviour which is different to that envisaged in the design.

Table 4.1 shows building classification in a convenient format.

4.3   Number of storeys

4.3.1   Introduction

The number of storeys a building has is fundamental to its classification. However, no definition of storey is given in AD-A$^{4.3}$. Although there is usually no dispute over the meaning, it has been defined elsewhere$^{4.5}$ as that part of a building which is situated between either:
– the top surface of two vertically adjacent floors
– the top surface of the uppermost floor and the surface covering of the building.

In the Scottish Regulations, it is 'the distance from the underside of one floor to the underside of the floor immediately above'.

Part-storeys are discussed in Sections 4.3.3 and 4.3.4.

4.3.2   Basements

Both AD-A$^{4.3}$ and BS EN 1991-1-7$^{4.4}$ are clear that if basements are designed to Class 2B, they do not need to be included as a storey in the storey count. The logic of this decision is obscure, but may derive from the greater reliability of horizontal ties anchored into perimeter walls. It follows that basements should only be treated in this way if all four perimeter walls (or equivalent structure) are present, and that part basements should otherwise be treated as an above ground storey.

The current regulations in Scotland and Northern Ireland include basements in the storey count but both permit exclusion of a basement from the count if it complies with the requirements of Class 2B.

An example of a part-basement is shown in the Copenhagen collapse (see Box 4.1). Reference 4.7

provides general guidance on the interpretation of AD-A Requirement A3 for many domestic and small buildings; it also amplifies the definition of a basement including part basements.

A helpful illustration of arriving at the number of storeys originating from NHBC has been published by SCI[4.8], and is reproduced in Figure 4.1.

| Box 4.1 | Copenhagen gas explosion |

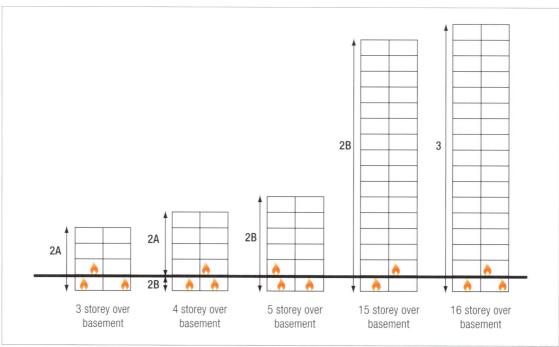

Before the collapse   After the collapse

The collapse of flats in Copenhagen following a gas explosion is instructive. The block was one of several on the site, with solid external masonry walls 350mm thick through the full height into which *in situ* reinforced concrete floor slabs were built at all levels. There were four floors of living accommodation over a semi-basement. This was set about 1.4m into the ground, so would qualify as basement by the NHBC guidelines[4.7]. However, from the photo it is difficult to see any reason for it to be discounted in assessing the building as being of five storeys.

The explosion occurred when gas from an external source leaked into the semi-basement and was very severe. The pressure lifted the floors thus relieving at least some, if not all, of the vertical load on the gable wall which then blew out. This left the floors above unsupported, resulting in the extensive collapse seen in the photo.

If horizontal and vertical ties had been provided, it is possible that enough of the structure would have remained intact to prevent the collapse.

Figure 4.1   Example of building classification

4.3.3   Part floors at roof level

It seems pedantic to classify a five storey building above ground wholly as Class 2B if the fifth storey is relatively insignificant. Clearly there is an area so small as not to increase the risk to the floors below enough to warrant designing the whole building as Class 2B. This amount is subjective, but might range from 25% for a building with a large floor plan (say $\geqslant 800m^2$) to 50% of a building containing only four flats (say $\leqslant 300m^2$) per floor.

4.3.4   Mezzanines and galleries

Similarly, it would appear reasonable to discount single level mezzanines and galleries satisfying the same limits to area given in Section 4.3.3.

4.4   Practical problems of interpretation

4.4.1   Introduction

Although the categories are quite precise, they should be interpreted with common sense. For example, having a floor area of $2001m^2$ does not automatically imply upwards classification. Conversely, having a floor area of $1999m^2$ implies care in considering classification.

4.4.2   Building or structure?

AD-A[4.3] refers to building classes, and uses the word 'building' throughout. Table 11, EN[4.4] also uses 'building' throughout, and replaces the AD-A term 'grandstand' with 'stadium'. SCI Publication 341[4.8] includes a diagram showing that a structure separated by a movement joint can be considered as two separate buildings. However, the wording of both AD-A and EN suggests that this is not what is intended (i.e. for downgrading classification) especially where an incident could affect both sides of the joint, and this view is reinforced by the example of the Paris airport terminal (see Box 4.2). Nevertheless, this principle should not be applied too rigidly: in a hospital with a number of single storey ward blocks grouped around a four storey block, it would be reasonable to treat each block as a separate building provided that each block was structurally independent and robust in its own right.

4.4.3   Ill defined uses

There are many building uses which are not explicitly covered by Table 11 of AD-A[4.3]. Nurseries and kindergartens probably have to be treated as educational, if (or as) the number of children per unit area is consistent. But what about doctors' surgeries and other day surgery units, care homes, hospices and other quasi hospitals? The appropriate tests are probably 'are people sleeping on the premises – and are they bedridden?' So doctor's surgeries and day surgery units can be treated as offices, care homes as residential but hospices as hospitals. AD-A (Section 5.4) refers readers to two reports which discuss the evaluation of risk and consequence and these could be considered in unusual cases.

The Institution of Structural Engineers Practical guide to structural robustness and disproportionate collapse in buildings   17

Box 4.2	Terminal at Paris Charles de Gaulle airport

Collapse of terminal building

The terminal building at Paris Charles de Gaulle airport partially collapsed in May 2004. It was 680m long and 32m wide, i.e. about $22000m^2$ per storey, and open throughout its length. If considered as a building, it would have come into Class 3, although at only three storeys it would have fallen outside the pre-2004 UK regulations. However, it was divided into ten separate structures each 68m long and with $2200m^2$ per storey. If this had been taken into account, it would only have come into Class 2B (regardless of whether the number of people was significant enough to take it into Class 3 in EN[4.4]).

It should be noted that the extent of collapse was limited by a movement joint at one end and a joint between precast concrete units at the other.

4.4.4   Multiple uses

The footnote to Table 11[4.3] states "for buildings intended for more than one type of use the class should be that pertaining to the most onerous type". Typical arrangements falling into this category are apartments over another use, such as parking, shops (which include restaurants), or a nursery. Another is where offices are combined with apartments or hotels. However, offices, hotels, apartments and other residential buildings all have the same cut-off levels, i.e. Class 2A up to 4 storeys and Class 2B up to 15 storeys, so there is no problem. With shops, the Class 2A limit comes down to 3 storeys but the Class 2B limit remains unchanged. If a nursery is treated as educational, then any building of more than 1 storey will fall into Class 2B.

4.4.5   Strong floors

If a strong floor can be designed to withstand collapse of the structure above, it clearly protects the occupants below. So a strong floor can be considered to be the foundation for the floors above and the number of storeys counted from this level up. However, below the strong floor the risk is unchanged, and the storey count should be that of the whole building. An example of the advantage of strong floors is the partial collapse of Torre Windsor in Madrid (see Box 4.3).

An application of this principle is a building comprising four storeys of say residential use on a podium at ground level; timber construction for the upper storeys is not unusual. Provided the podium is designed to meet the Class 2B requirements and

Box 4.3	A strong floor located within a tall building's height can contain debris from collapses above

Partial collapse of Torre Windsor office building in Madrid

The 30 storey Torre Windsor office building was constructed in the 1970s. It consisted of a reinforced concrete core and six reinforced concrete columns within the floor plate area, and steel load-bearing mullions (steel edge columns) around the perimeter. At the time of design, the relevant codes did not require these mullions to have any fire protection. The floor was of concrete waffle slab construction. There were strong transfer floors at the 3rd and 17th levels.

During refurbishment work, a fire broke out on the 21st floor. The fire spread downwards to the 2nd floor, and upwards to the top of the building. In the absence of any protection, the mullions weakened in the heat. A sufficient number lost their required load capacity causing sections of the building above the upper strong floor at level 17 to collapse. It is likely that only the presence of this floor prevented total progressive collapse [4.9].

not to collapse in the event of collapse of the storeys above, the storeys above can be designed to meet Class 2A. The floor forming the roof of the podium would need to be designed to carry the debris from the collapse, but it would not be unreasonable to take this as a static load spread over an area up to say 25% larger than the original footprint. Although the fall of debris is clearly a dynamic event, in the notional method of design adopted, the presumed loading is simply the self weight of the debris.

4.4.6   Acoustics

A particular problem arises in taller blocks of flats which have to be designed for both sound insulation and robustness and the requirements for each are conflicting. For reasons of acoustic isolation, tying of units between flats is not normally permitted. This is an area of difficulty for which little guidance is available. Until further guidance becomes available, it is recommended that advice is sought from unit suppliers who may have specific test data available.

4.5   Extensions, alterations and change of use

4.5.1   Introduction

A significant problem faced by designers is determination of class when a building is being extended, altered or undergoing a change of use. It is easiest to consider these in reverse order. The discussion is based on the regulations [4.1] applicable in England and Wales, but the requirements in the other parts of the UK are similar (but see Section 4.5.5 in respect of Scotland).

4.5.2   Change of use

Change of use is covered by Regulations 5 and 6 [4.1]. Requirement A3 only applies where the change of use is to: a hotel or boarding house; an institution (i.e. accommodation for elderly or disabled people or under-fives); a public building (e.g. a theatre, public library, hall, place of worship, school or educational establishment); or a building previously exempt (e.g. a temporary building). Note that changes to residential use or to offices or shops are not included. The only changes that appear likely to trigger the need for a re-appraisal to A3 are: a building between two and four storeys converted to education; or a building of three or more storeys, or over 2000m² per storey, converted to public admission. Nevertheless the general caution about the unusual applies and class upgrading should be considered whenever the use change significantly increases public risk.

4.5.3   Alterations

Alterations are covered in Regulation 3(2) [4.1], where an alteration has to be taken into account only if it is material. A material alteration is one that results in either: a compliant building becoming non-compliant; or a non-compliant building becoming more unsatisfactory in relation to the requirements. An example of an alteration to a building without extending it could be by forming a light well. Then it might be argued that the building will be no more unsatisfactory than before, thus that particular alteration is not material, and that therefore no work is required to meet A3. Another example is converting a single family house into flats. Again it could be argued that the building will be no more unsatisfactory than before, but inserting new stairs or a lift might affect compliance.

4.5.4   Extensions

Extensions are covered by being included in 'building work' (see Regulation 3(1) [4.1]). This appears to mean that any extension, whether sideways, upwards (see Figure 4.2) or even downwards has to meet requirement A3. However, there is no requirement to apply A3 to the original building retrospectively unless it is altered (then see above). Neither is it necessary to separate new from old with a movement joint, although it will often be expedient. While this makes sense for sideways extensions, it is apparent that putting a penthouse floor on the roof might make the original building more unsatisfactory. The same argument applies to forming a new basement. This aspect is discussed further below.

Figure 4.2   Typical extension upwards on an existing block of flats

The limited objective of being 'no more unsatisfactory than before' is helpful, as it implies that while new buildings should meet or even exceed the regulations, it should be possible to tolerate a degree of compromise with existing building stock. How this is achieved must be argued on a case-by-case basis, but it should be remembered that while the Building Regulations[4.1] are mandatory, AD-A[4.3] is advisory. Constructing a lightweight fifth penthouse floor on top of a reasonably sound existing four storey building would appear to maintain the risk close to the original level, and thus to meet the 'no more unsatisfactory' requirement in principle even if not precisely. It would also appear to 'secure reasonable standards of health and safety' (see Section 3.2 re: Regulation 8[4.1]) and could perhaps be described as 'only marginally more unsatisfactory than before'.

4.5.5   Situation in Scotland

The Building (Scotland) Regulations 2004[4.10] (Schedule 2 to Regulation 4) designates a specific category of changes in the use or occupation of buildings as being 'Conversions'. Regulation 12 requires that when the use change falls within the category of a conversion then the converted building must be altered or strengthened to the standard required by current structural regulations in so far "…as is reasonably practicable, and in no case be worse than before the conversion…". Reasonably practicable is defined in this case as "…having regard to all the circumstances including the expense involved in carrying out the work."

This is a stiffer test than that described above for buildings in England and Wales, but the philosophy outlined should be applicable to both jurisdictions.

4.5.6   The Camden ruling

An approach to resolving this dilemma that has been followed in the past in the UK is that offered by the London District Surveyor's Association internal document *Guidance on achieving compliance on disproportionate collapse in existing buildings for Class 2B cases in single/multiple occupancy* (available to local authorities) previously known as the Camden ruling. This suggests adopting a design that demonstrates that any damage occurring within a fifth storey would be contained by the floor forming the roof to the fourth storey (including its own supporting structure). If this can be done, the alteration appears not to change the risk to the occupants of the original lower four floors and the structure could thereby be assessed as Class 2A rather than Class 2B.

However, this approach usually requires the original roof to be strengthened or even a new strong floor to be inserted above it. Therefore if a collapse were to occur in the lower floors, the weight of the resulting falling debris would be considerably greater. For this reason, the 'only marginally more unsatisfactory than before' argument proposed above is preferred to the Camden ruling.

Another problem occurs when adding a storey to a building of 2 or 3 storeys built before 2004. The original construction would have fallen outside Regulation A3 (and now equivalent to Class 1), while the proposed extension brings it into the definition of Class 2A. Again this must be argued on its merits, but the presumption of encouraging sustainable development would suggest that the 'only marginally more unsatisfactory than before' argument might be applied here also.

4.5.7   General

Total and partial collapses have occurred during refurbishment so especial care is required when working on older non-framed buildings. General guidance may be obtained from Reference 4.11.

Generally, when contemplating any extension onto an existing building or any alteration, engineers may consider a risk assessment process to demonstrate robustness. A holistic approach is important, considering existing features such as returns on walls, chimney breasts, and any other features that enhance structural robustness. Large openings and previous alterations may have introduced weaknesses. An example of the process developed by LDSA/LABC is included in Appendix 1. The assessment process should always include consideration of what might happen to adjacent structures and whether the proposed alterations have in any way degraded their robustness.

4.6   References

4.1   *The Building Regulations 2000*. London: The Stationery Office, 2000 (SI 2000/2531), as amended by *The Building (Amendment) Regulations 2004*. [s.l.]: The Stationery Office, 2004 (SI 2004/1465)

4.2   Ministry of Housing and Local Government. *Report of the inquiry into the collapse of flats at Ronan Point, Canning Town*. London: HMSO, 1968

4.3   Office of the Deputy Prime Minister. *The Building Regulations 2000. Approved Document A: Structure*. London: NBS, 2004

4.4   *BS EN 1991-1-7: 2006: Eurocode 1: Actions on structures – Part 1-7: General actions – Accidental actions*. London: BSI, 2006 and *NA to BS EN 1991-1-7: 2006: UK National Annex to Eurocode 1 - Actions on structures – Part 1-7: Accidental actions*. London: BSI, 2008

4.5   Brick Development Association et al. *Masonry design for disproportionate collapse under Regulation A3 of the Building Regulations (England and Wales)*. Available at: http://www.brick.org.uk/_resources/Masonry%20 Design%20for%20Disproportionate%20Collapse%20 Requirements.pdf [Accessed: 1 February 2010]

4.6   Scottish Building Standards Agency. *The Scottish building standards technical handbook: non-domestic*. Edinburgh: The Stationery Office, 2010

4.7   National House Building Council. *The Building Regulations 2004 edition – England and Wales: Requirement A3 – disproportionate collapse*. Available at: http://www.nhbc.co.uk/NHBCpublications/ LiteratureLibrary/Technical/filedownload,23676,en.pdf [Accessed: 1 February 2010]

4.8   Way. A.J.G. *Guidance on meeting the robustness requirements in Approved Document A. SCI Publication P341*. Ascot: SCI, 2005

4.9   Standing Committee on Structural Safety. *The Fire at the Torre Windsor Office Building, Madrid 2005. SCOSS Failure Data Sheet SC/08/024*. Available at: http://www.scoss.org.uk/publications/rtf/ SC08024FireatTorrWindsorbuildingMadridMay08.doc [Accessed: 1 February 2010]

4.10   *The Building (Scotland) Regulations 2004*. Edinburgh: The Stationery Office, 2004 (SSI 2004/406)

4.11   Institution of Structural Engineers. *Appraisal of Existing Structures*. 3rd ed. London: Institution of Structural Engineers [due 2010]

5 Designing for robustness

5.1 Introduction

This chapter describes approaches required to achieve a robust structure. The principles discussed are applicable to all materials, though the manner of application is often material dependent and so specific advice is given in the relevant material chapters. The importance of considering robustness from the earliest stages of a project is highlighted followed by the need for clear direction to ensure the philosophy developed is implemented within the final design, detailing and construction.

For most structures, the standard means of achieving robustness is by compliance with the rules in Approved Document A[5.1] and BS EN 1991-1-7[5.2]. From these documents, the design criteria for the various classes of structure are as follows:
- Class 1 - No additional measures (i.e. other than ultimate and serviceability design).
- Class 2A - Provide effective horizontal ties or effective anchorage of suspended floors and roofs to walls, as described in the materials codes. BS EN-1991-1-7 states that for this class of structure, horizontal ties should be used for frame structures, and anchorage of suspended floors and roofs should be adopted for loadbearing wall construction. In some cases, it may also be appropriate to adopt horizontal tying for loadbearing wall construction.
- Class 2B - Provide effective horizontal ties to floors and roofs plus effective vertical ties or apply notional column/wall removal or design as key elements (explained below).

Note the difference in the treatment of horizontal ties between 2A and 2B: compliance with Category 2A may be achieved by 'effective anchorage of suspended floors to walls' rather than 'provision of effective horizontal ties'.

While the intentions of the categories are clear enough, buildings come in a variety of forms and interpretation is required especially in structures of mixed form (hybrid structures) or where alterations are being made.

A particular problem arises out of possible interpretations of AD-A[5.1] versions. Notwithstanding this approach whereby horizontal ties are not specified as a requirement when elements are removed or designed as key elements, it is the Task Group's view that provision of ties is always required in Class 2B buildings unless there is good evidence to the contrary. Good engineering requires horizontal linkage across the structure though there is a question of magnitude and form to be resolved which will be material specific.

In some industries, it has not been the practice to provide any additional physical ties if the structure is proven robust enough via the element removal approach. An example of this would be that pursued in timber engineering where full scale tests on timber

framed structures (see Chapter 9) have proven Class 2B structures adequate provided certain detailing rules for floor anchorage are adopted. This process is reflected in **Note b** of the flow chart of Figure 5.1

Figure 5.1 provides a flow chart to guide the user through the various robustness options for the building classes defined above. Although Class 3 buildings are not considered in this *Guide*, there is a widely held view that they should meet the requirements for Class 2B as a minimum standard, in which case the explanation of design to Class 2B in this *Guide* will be relevant.

Although the guidance in this document is predominantly concerned with achieving structural robustness within individual buildings, opportunities to improve overall robustness by protecting the structure from hazardous events should always be considered when these events are reasonably foreseeable, e.g. protection of columns from vehicle impact, ensuring gas supplies are separated or protected etc.

Robustness of a structure itself can be considered at two levels, the overall structural concept and then detailed provisions. These are discussed below.

5.2 Structural concept

A building's structural form will significantly affect its robustness (this is discussed in more detail in Chapter 2). Traditional cellular forms with many loadbearing walls assure a sensible level of robustness because loss of any one wall will generally not lead to the collapse of a large proportion of the structure. In contrast, having a large span supported on an easily dislodged and/or vulnerable single column would not be a robust structure. At the concept stage, the layout of the building and its basic structural action will need developing. For any structure, there should be an assessment of the hazards and provision of clear paths for horizontal and vertical loads back to the foundations. Additionally, a robust structural concept will be one which avoids situations where damage to small areas or failure of any single element progresses to widespread collapse. Notwithstanding that ideal, there are clearly occasions when reliance does have to be placed on single elements, but at least once this is recognised, their robustness can be improved by making such elements substantial. Experience has shown certain arrangements to be potentially vulnerable; examples include:
- significant transfer beams, these are single beams which support a number of columns or hangers
- apparently minor elements that are required to ensure the stability of more significant elements
- significant cantilevers
- long span, simply supported beams.

The last two forms have no redundancy but that need not imply unacceptable vulnerability.

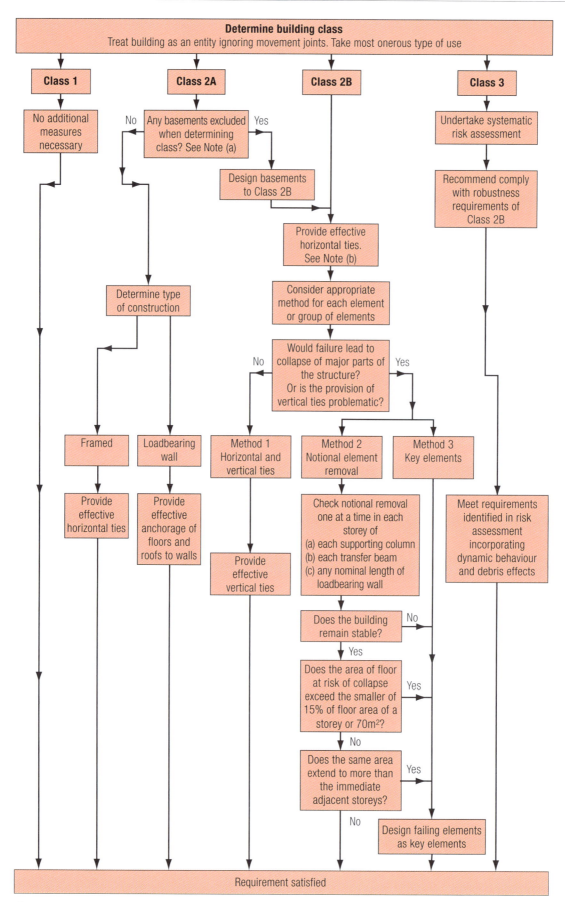

Figure 5.1   Disproportionate collapse – flow chart

Notes

a   Rules on the exclusion of basements vary in the building regulations of the various regions. When designing to EN1991-1-7[5.2] basements are included in the storey count. However in England and Wales and Scotland and Northern Ireland, if they satisfy the Class 2B requirement, they may be excluded from the count.

b   In Class 2B, for horizontal ties, rather than using physical ties in some cases alternative methods may be demonstrated by test.

Some judgement is required on whether the element considered is significant; for example failure of a transfer beam supporting a single storey at high level is not as critical as one supporting a whole building façade. However, functional restrictions on structural form may limit a designer's scope for avoiding arrangements that are potentially non-robust. Nonetheless, early identification will allow development of a strategy to provide the required compensatory robustness within the element itself. Once highlighted, a significant transfer beam could be designed to withstand a certain event (see key elements below) or redundancy could be built into bracing systems such that the loss of a set number of braces did not permit restrained elements to buckle or the structure as a whole to be placed at risk.

During project planning, it should be acknowledged that element detailed design may be carried out by someone other than the lead designer. This does not remove the need for the lead designer to take responsibility for design compatibility of the various elements including responsibility for overall robustness (see Section 2.13). As a minimum, the overall robustness concept for the building, and therefore the effect that this will have on the design of individual elements, should be documented for later phases of the design development (which may be carried out by others). On completion of the design, the structure should be checked to confirm that the initial assumptions are satisfied. This may include confirming that loads and load paths remain as envisaged and that the overall robustness concept has been achieved.

5.3 Notional horizontal loads

All structures should be capable of carrying horizontal loads applied in any direction on a horizontal plane and there should be a defined load path for such loads back to the foundations. Wind action will generally suffice as a sufficient load to assess stability. However in some structures, that loading is absent or not obvious (see Section 2.4) and in others it may be insufficient.

The differing historical development of notional horizontal loads leads to some detailed differences in application. In the UK, the original concept was to ensure that low rise buildings which had otherwise very low wind loads applied were, nonetheless, designed to have some capacity to resist accidental horizontal loads. However, in Eurocodes such as BS EN 1992-1-1[5.3] the horizontal load is developed from a notional out-of-verticality. As such, the horizontal load is a function of vertical load and always present whenever vertical load is present. Consequently in BS EN 1992-1-1, the horizontal loads due to this notional out-of-verticality are added to the wind loads in accordance with the appropriate load combination. Likewise in UK practice, the load combinations considered do include a percentage of wind loading in the accidental load case combination. At first sight this seems improbable since accidental loading is unlikely to coincide with high wind loading but if the notional loading is related to out-of-verticality, its inclusion becomes more rational. Whilst the codes differ in approach, the effect in all cases is to ensure structures are robust against the effects of construction tolerance deviations, subsequent settlements or accidental horizontal loads.

5.4 Detailed provisions

Detailed provisions for robustness are given in material specific codes of practice and described in later chapters but the principles are described below. There are typically three approaches:
- **Approach 1:** The indirect design method: provision of horizontal and vertical ties.
- **Approach 2:** Alternative load path method: notional removal of elements.
- **Approach 3:** Specific load resistance method: the provision of key elements.

It is emphasised that these separate approaches are largely based on judgement as providing a level of robustness commensurate with routine risks and are achievable at affordable cost. It is not difficult to identify mathematical inconsistencies in them but the sufficiency of the proposals has been proven over time. Potential accidental actions and even building forms change over time and their relevance to any particular design should be considered.

In Approach 1, applicable to Class 2A and Class 2B buildings, resistance to progressive collapse is considered implicitly 'through provision of minimum levels of strength, continuity and ductility' throughout the whole structure (this is sometimes called the Indirect Design Method). Adopting this method should provide buildings with sufficient robustness to survive a reasonable range of undefined accidental actions. The tying provisions are more onerous in Class 2B buildings than for Class 2A buildings as a reflection of the potentially greater risks. Tie capacity is typically provided by the structural members themselves but also by making sure that their connections or anchorages are strong enough (this implies also that any walls used for anchorage are strong enough. Particular care is needed say when anchoring to bricks bedded in lime mortar). This is primarily achieved by design using conventional procedures to carry defined tie forces.

In Approach 2, the alternative path method presumes that through abnormal loading a critical element is removed and the structure is thereafter required to redistribute its gravity loads to the remaining structural elements via alternative load paths. There is no requirement in the UK to consider dynamic loads associated with the element removal. In practice, elements are notionally removed one by one and the residual structure (members and connections) tested for strength. Local collapse is not prohibited but its extent must not exceed prescribed limits.

In Approach 3, certain elements are designed to sustain a notional upper bound to the abnormal loading (a value of 34kN/m^2 imposed pressure is used derived from the Ronan Point blast) albeit on a just survive basis; the presumption then being that such members are strong enough to cope with a range of events.

It should be noted that Approaches 2 and 3 are principally concerned with vertical structure or elements supporting vertical structure. When applying these approaches the designer must still ensure that the horizontal structure is robust in both directions. This is generally achieved by providing horizontal ties.

The application of these approaches is discussed below.

5.5   Tying

The provision of ties having a defined capacity and linking components helps to constrain the elements from displacement during an event and can make possible alternative load carrying systems including catenary and vierendeel action.

Roofs need to be tied down even in single storey buildings (AD-A[5.1] P29 and Diagram 16) not least because of the vulnerability of light roofs to dislodgement in wind suction; but designers need to consider what is appropriate for the particular circumstances.

5.6   Horizontal ties

The demands for horizontal ties differ between Classes 2A and 2B. In Class 2A, it is possible to apply effective anchorage of suspended floors to walls. In Class 2B buildings, cross ties should be provided. Reference is then made to the various material codes of practice which generally split the horizontal ties into two categories. Ties around the structure are called peripheral ties whilst ties across the structure are internal ties. Along a particular load path (which must be continuous) different structural elements (say a series of beams) may be used as the ties, providing they have adequate interconnection. Rules for tie location and for their design forces are given in the material codes (BS and EN).

Ties can act to prevent the structure being dislodged which is particularly important when supports are narrow, or at a perimeter where the ties must be capable of resisting any outward force on the supporting vertical element. The need for such tying was demonstrated in the Ronan Point collapse where a gas explosion blew out a loadbearing wall, causing the slabs above to collapse. Lack of tying was also a factor in the Camden School collapse[5.4]. Chapter 4 suggests that tying might have limited the collapse extent in the Copenhagen blast (see description in Box 4.1).

Ties need to be continuous (i.e. lapped or connected) across from edge to edge or around the structure, while at their ends, horizontal ties to edge columns and walls must be satisfactorily anchored back. All tie force paths should be geometrically straight; changes in direction to accommodate openings or similar discontinuities should be avoided wherever possible. Where such changes are unavoidable, the tendency of the tie to straighten under load should be considered and restraining elements provided. For buildings composed of separate structures, or incorporating joints creating structurally independent sections, the tie force requirements are applied to each independent section, each treated as a separate unit.

The code specified tie forces aim to ensure that beams or slabs can span across a removed support. However, there is no theoretical justification that ties designed to the codes will in fact enable the structure to span across a damaged area in all possible circumstances. Indeed a number of theoretical objections can be raised and a particular difficulty exists when trying to justify the sufficiency of ties following removal of a corner column.

One further difficulty in relying on the benefits of either catenary (see Section 5.9) or membrane action lies with justifying the significant ductility required. Rules for tie design and location are given in the material codes but there are no direct requirements for providing a ductility magnitude, this being assumed. Comments on providing appropriate ductility are given in later materials chapters (Chapter 2 also discusses ductility and energy absorption).

The rationale for ignoring all these objections to the nominal regulation rules is that a balance has to be drawn between the risks of an event occurring on the one hand against the cost of tie provisions on the other. The implication of the AD-A[5.1] is that a notional tie provision represents a reasonable level of precautionary investment. The consensus is that horizontal ties are an important safeguard which should always be incorporated and will safeguard most buildings for most hazards. But outside that usage, their potential weaknesses should be accounted for more directly, especially for more demanding structures.

5.7   Vertical ties

Vertical ties have two roles. The first is to provide some form of minimum resistance to vertical elements being removed. The second is to enable load sharing between floors above a damaged vertical element. By linking a number of floors together, it is possible to provide a load path back to intact structure above, perhaps by developing vierendeel action.

The rule for vertical ties (see Chapters 6 to 10) is that each column or wall should be able to support the largest dead and imposed load reaction applied to the column or wall from any one storey (above or below). The requirement for the largest load could be problematic to define, and BS 5950[5.5] for example limits this so that each column splice only has to carry the largest load arising between it and the next splice down. In practice, if it is possible to provide vertical ties at all, their capacity is not usually a problem.

When required, vertical ties must be continuous from the lowest level to the highest level and this includes anchorage into the footing or foundation. The rationale for tying into foundations is that this helps reduce the possibility of lowest column removal. Where such ties into the basement are not practical, consideration as a key element (see below) normally provides a practicable solution.

5.8   Element removal

An alternative to the tying method of providing robustness is to consider notional element removal. The term 'notional' is used deliberately as a means of emphasising an imaginary scenario.

In this approach, a defined element is removed (the element might be a column or nominal wall length) and the consequence examined. It is assumed that the act of removal itself does not induce any forces, static or dynamic. Equally no benefit should be taken for enhanced material strength due to fast strain rate effects. After each removal, the building as a whole must remain stable and its members must not be locally destabilised by the removal, save for acceptance of localised damage within the prescribed limits. If such removal cannot be tolerated, the member must be designed as a key element. AD-A[5.1] and BS EN 1991-1-7[5.2] give guidance on elements that ought to be removed and on the corresponding maximum area of collapse permitted.

What has to be hypothetically removed one at a time in each storey is "each supporting column and each beam supporting one or more columns (usually called a transfer beam) or any nominal length of loadbearing wall" (see Figures 5.2 and 5.3). Loadbearing construction includes masonry walls and walls of close centred timber or light steel studs. The nominal length removed is generally 2.25H where H is the storey height (or the clear height in the Eurocode) but in the case of an external masonry, timber or steel stud wall, the length removed is that between "vertical lateral supports" (usually wind posts or return walls). For loadbearing walls at the corner, the removed length of wall should be equal to the wall height in each direction but not less than the distance between expansion or control joints.

Critical cases for wall removal often occur where the plan geometry of the structure changes significantly such as at abrupt decreases in bay size or at re-entrant corners, or at locations where adjacent columns/walls are lightly loaded, where the bays have different tributary sizes, and where members frame in at different orientations or elevations. Provided consideration is given to this approach in the design phase, it is usually possible to show that the unsupported structure can survive via some alternative structural system.

A weakness of AD-A[5.1] is a case where support is provided by a number of closely-spaced columns. It would be irrational just to remove a single column, and the recommendation here is that all columns within a plan diameter 2.25H should be removed simultaneously.

An alternative path analysis should be examined for each floor, one at a time. For example, if a corner column/wall section is specified as the removed element, one analysis is performed for removal of the ground floor corner column/wall section; another is performed for the removal of the first floor corner column; another alternative load path analysis is performed for the second floor corner column and so on. If the designer can show that a similar structural response is expected for column/wall removal on multiple floors, then the analysis for these floors can be omitted but the justification for not performing these analyses should be documented.

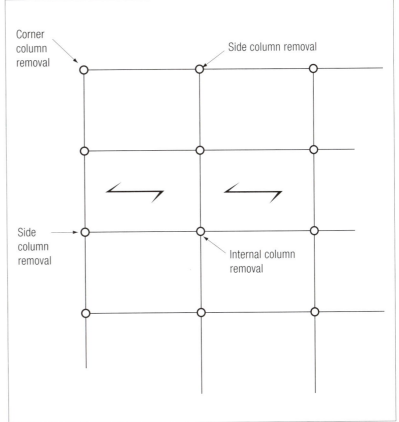

Figure 5.2   Location of column removal for alternative path method

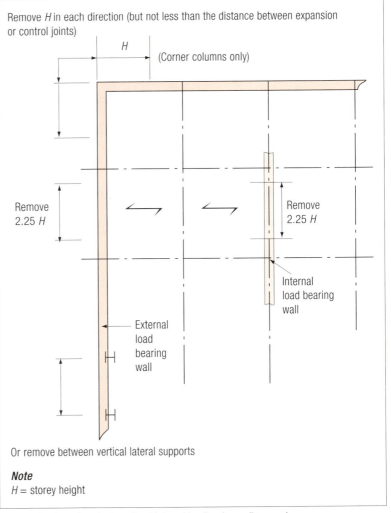

Figure 5.3   Location of external and internal loadbearing wall removal

Following element removal, strategies for survival might be:
- Spanning the floor at right angles to its normal design case or utilising two way spanning (in timber this might make use of the floor boarding).
- Making use of the available reinforcement in reinforced concrete slabs and beams to enable the element to span two bays, albeit not necessarily by pure bending action.
- Using the ability of a wall to cantilever as a deep beam over the notional opening (and this may be required where there is loss of a corner support). In a similar way, using the wall above as a deep beam to span over an opening.
- The provision of bridging elements such as those used in timber platform frame construction (see Chapter 9).

When considering alternative load paths after element removal, all the forces required should be accounted for. For example if catenary action is assumed, the horizontal forces developed must be resisted by the remaining structure.

5.9   Catenary action: horizontal and vertical

After element removal, a key survival strategy is to rely on the ability of the remaining connected members to span via catenary action accepting that deflections may be high. The strategy may rely on pure catenary action where the tension steel takes axial load only. But more realistically, slabs and beams will retain some degraded moment capacity at their ends and centre which will reduce the tie forces demanded. There is also a common case where the catenary receives some midspan support via tie forces from a surviving column above (see Figure 5.4). Because of the complexities in all this, calculations cannot be accurate; they can only be crude approximations to suggest survival or failure. Currently there are no further guidelines and designers must assess individual cases on their structural merits.

The theoretical benefits of catenary action can be seen from Figure 5.5 (and Figure 5.4) which shows a member carrying a distributed load w albeit with significant sag deformation. The anchor force required R_h is $wl^2/8h$, but h is indeterminate. If calculation is required, equilibrium can be established fairly quickly by trial and error: the maximum credible value of R_h is the ultimate strength of the member as a tie (or some lower connection capacity) and the maximum credible value of h can be evaluated from the plastic strained length of the tie or a sag based on some presumed joint articulation. The maximum realistic value of h before total failure will be material dependent. If the value of R_h so determined exceeds $wl^2/8h$, there exists some lower value of R_h which will suffice to support the structure (where there is some residual moment capacity, use of $wl^2/10$ or 12 might be more appropriate than $wl^2/8$).

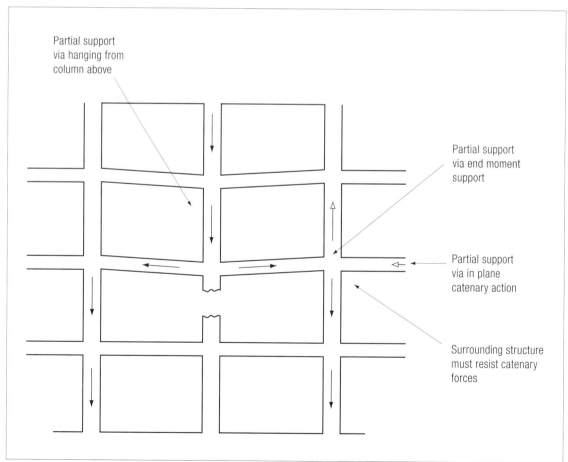

Figure 5.4   Floor survival via catenary and hanger action

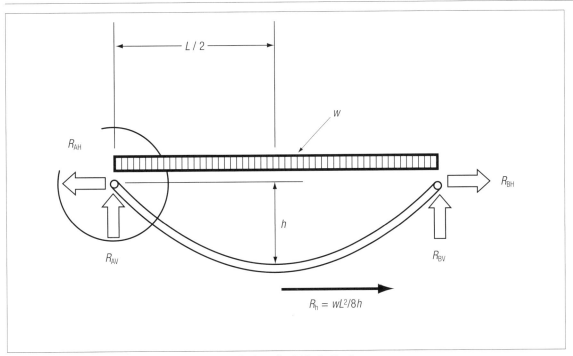

Figure 5.5   Analysis of a tension member supporting a horizontally distributed load

In two-way spanning structures, similar benefits can be achieved via radial tensile membrane action and indeed when considering the loss of an internal column below a slab (or equivalent) it is possible to develop an in-plane compression ring action to resist the radial tie forces generated[5.6]. Figure 5.6 shows such a slab structure with new equilibrium obtained, not reliant on bending action alone.

Clearly catenary systems only work if there is enough ductility in the joints to form and sustain a mechanism and if R_h can be anchored; the structure surrounding the tie end must also resist R_h. Where support from a column hanger is assumed, the structure around the hanger top must be capable of sustaining the assumed capacity. An appropriate value of w along with appropriate factors is discussed in Section 5.13.

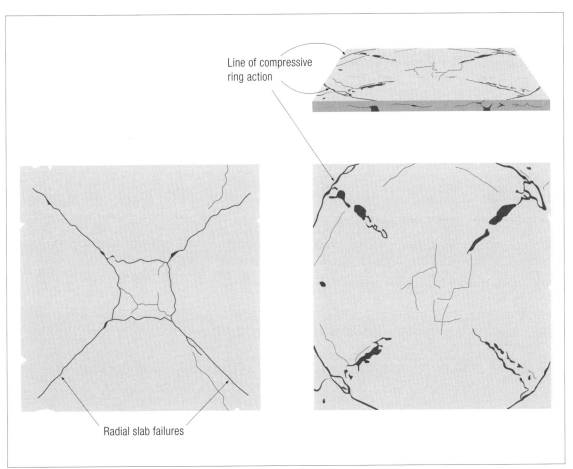

Figure 5.6   Tensile membrane action (test results from a concrete slab)

5.10   Partial collapse and debris loading

Where the structure cannot support damage by the means described at the end of Section 5.8 or in Section 5.9, limited collapse is permitted. This is explained in AD-A[5.1] as "the area of floor at any storey at risk of collapse does not exceed 15% of the floor area of that storey or 70m², whichever is smaller, and does not extend further than the immediate adjacent storeys". Figure 5.7 shows the concept. BS EN 1991-1-7[5.2] increases the area to 100m² and clarifies the limit as "two adjacent storeys'. The permissible area of floor collapse is empirical and it is likely that the next edition of AD-A will be changed to be consistent with BS EN 1991-1-7. (The current Scottish regulations already refer to the Eurocode area definition.)

The limitation on storey spread implies that the floor below must be able to support debris accumulation. Dependent on the characteristics of the presumed collapse, the debris load may be less than the full load of the collapsed part as illustrated in Figure 5.8. However as the collapse of two adjacent floors is allowed, the debris loading may be that from two floors. It is not possible to be specific, each case must be argued on its merits. But, however this is done an allowance for imposed load must be included, consistent with the accidental load case (see Section 5.13) and with the most likely failure mechanism (see also Section 4.4.5).

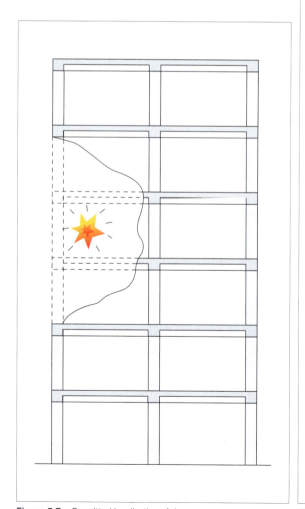

Figure 5.7   Permitted localisation of damage (two adjacent floors)

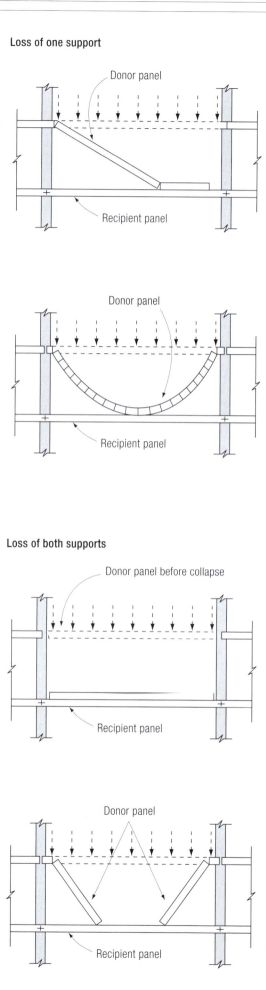

Figure 5.8   Possible collapse modes of floors and roofs

5.11 Key elements

A third approach, which is often easier to consider than element removal, is to consider members as key elements, in effect making them strong enough to withstand a prescribed hazard loading. The approach is beneficial if intuitively such members are strong enough to survive. The key element approach offers advantages if loss of that single member would otherwise lead to loss of a significant portion of building; examples include loss of a significant transfer beam or a column supporting such a beam.

The general design approach for key elements is to consider uniform pressure acting over their surface (in each orthogonal direction, one direction at a time) plus the surface of any attached items such as cladding. Likewise, any member restraining a key element should be designed for the specified design pressure or the restraining element benefits ignored. The restraining element can be checked separately to the key element however, if both would be affected by the same event, then it is more logical to consider both elements loaded simultaneously. The pressure is applied in conjunction with the accidental load case (see Section 5.13).

A pressure value of 34kN/m² (derived from Ronan Point) is used as the test design pressure representing the static equivalent pressure from a notional gas explosion. The force transferred from the attached item may be reduced to a realistic estimate governed by an upper bound of the fixing capacity. It would be unduly onerous to apply the accidental design pressure over large areas, e.g. to slabs attached to (and stabilising) long span transfer beams. The 2.25H limit on length for loadbearing walls usually corresponds to no more than 6m, so applying the pressure to an area limited to 6m × 6m would seem reasonable, although any such relaxation should be considered in light of the specific circumstances.

The pressure of 34kN/m² is onerous for walls and slabs but often non-critical for isolated columns due to their limited surface area. But structural sense should be used; the value of lateral load generated on an unusually narrow column may not be an appropriate design force. So for all key elements, the possibility of other accident scenarios should always be borne in mind. Alexander[5.7] has proposed a static equivalent impact load of 150kN for columns increasing to 250kN for the ground storey. These values are reasonable for typical situations but again the actual load used should be justified on a case by case basis. For buildings adjacent to highways further guidance on vehicle impact is given in BS EN 1991-1-7[5.2].

5.12 Transfer beams

The consequences of any transfer beam failure (see Figure 5.9) or of its supporting structure is likely to be more serious than the failure of normal beams or columns and this was well illustrated in the major failures at Oklahoma City[5.8, 5.9] (see Box 5.1). Not all transfer beams support such large amounts of structure; a worst case probably arises when the transfer beam is on the building perimeter with columns above and hangers below. Where transfer beams are clearly carrying significant portions of a building, the standard tie forces could prove inadequate, and the recommended approach is to design the beams and their associated structure to:
– only sustain localised damage, or
– be removable elements, or
– be designed as key elements.

Box 5.1	Murrah building, Oklahoma City

End of transfer beam that was destroyed

Blast damage and collapse of the Murrah Building, Oklahoma City

The Murrah building had a reinforced concrete frame and was stabilised by concrete shear walls. At the third floor, the nine storey building had a transfer beam on the outside with the column spacing above half that of below. In 1995, the building was truck bombed at the front and the blast destroyed one of the lower columns, two adjacent columns failed in shear. This left the transfer beam spanning about 50m and so it failed taking with it 50% of the building floor area. Most of the failure was due to progressive collapse and it all occurred within about 3 seconds.

There was no continuity steel in the transfer beam over intermediate column supports so in effect, after column removal, it was a beam without bottom reinforcement and gross failure was inevitable.

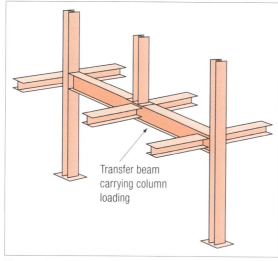

Figure 5.9   Transfer beams

Usually the recommended loading and material factors will give practicable solutions in design calculations. But if problems arise, particularly in the assessment of existing structures, designers should use common sense to consider an appropriate probability of accidental loading using best estimate values of loading and material properties. The emphasis on achieving robustness should always be sound logic and structural principles, not just on artificial numerical compliance with a code.

5.14   Summary

Three principal methods of providing robustness have been described. The use of ties is event independent, and the evidence is that within limits of applicability, ties provide adequately robust structures preventing disproportionate collapse in real events. Tie effectiveness is probably limited by element ductility or more likely joint capacity/ductility. For this reason, extrapolation of tying rules beyond the situations provided for in current guidance is not recommended.

Both the element removal and key element approaches are implicitly limited to proving adequate protection against a prescribed design event (i.e. the one which removes the element) and they may not therefore provide protection for bigger event magnitudes. Indeed for element removal, it is reasonable to accept that failure areas larger than the limits defined as acceptable in the guidance are not actually disproportionate if the event causing the failure is extraordinary. Nonetheless, the prescribed events restricted to single element removal are well established in current guidance and appear to provide adequate protection for the majority of structures.

5.13   Design load cases

Calculations are required when assessing structures for imposed accidental loading. Two issues arise: what the appropriate load factor should be; and what are the likely co-existing loads. Overall, in the improbable event of an accident or explosion, the structure can be pushed close to its ultimate capacity so a low load factor is justified. Moreover, for most structures, it is unlikely that the event will occur when the structure is carrying its full imposed or full wind loads. It is even less likely that high imposed loading and high wind load will coincide during the event. In the Eurocodes, the accidental load combination is given in BS EN 1990[5.10] and the relevant National Annex. For UK combinations see Box 5.2.

It can be seen that the approaches are broadly compatible but that the Eurocode approach is likely to give lower loads where wind is dominant. This is sensible as it is very unlikely that even a third of the wind load will be present at the same time as the accidental action. However as noted in Section 5.3, under the Eurocode system, all vertical loads have a horizontal component which should be added to any wind load in the load case.

The material codes also allow the reduction of some material safety factors during an accidental loading. The relevant material sections provide further details.

Whichever approach is adopted, the robustness of the structure will be improved by careful detailing to assure ductility and energy absorption and the other attributes outlined in Chapter 2.

The sequence of design for robustness starts with choosing a structural layout that limits, or avoids, the use of any elements whose failure would lead to collapse of a significant part of the structure. Thereafter the design process is as summarised in Figure 5.1.

Box 5.2	Load combinations
1.0 permanent + 1.0 accidental + 1.0 frequent value lead variable + 1.0 quasi permanent value of other variable actions.	
Where a variable action has beneficial effects, it is ignored completely.	
For a typical office building this would lead to the following factors:	
= 1.0 permanent + 1.0 accidental + (zero or) 0.5 imposed and 0.0 wind, with the floor load as the lead variable action, or	
= 1.0 permanent + 1.0 accidental + (zero or) 0.3 imposed and 0.2 wind, with the wind load as the lead variable action.	
This can be compared to the accidental design load for an office building in BS 8110 (Part 1)[5.11] of: = 1.05 (1.0 dead + 1.0 accidental + 0.33 (or zero) imposed + 0.33 wind)	

5.15 References

5.1 Office of the Deputy Prime Minister. *The Building Regulations 2000. Approved Document A: Structure.* London: NBS, 2004

5.2 *BS EN 1991-1-7: 2006: Eurocode 1: Actions on structures – Part 1-7: General actions – Accidental actions.* London: BSI, 2006 and *NA to BS EN 1991-1-7: 2006: UK National Annex to Eurocode 1 - Actions on structures – Part 1-7: Accidental actions.* London: BSI, 2008

5.3 *BS EN 1992-1-1: 2004: Eurocode 2: Design of concrete structures – Part 1-1: General rules and rules for buildings.* London: BSI, 2004

5.4 Department of Education and Science. *Report on the collapse of the roof of the assembly hall of the Camden School for Girls.* London: HMSO, 1973

5.5 *BS 5950-1: 2000: Structural use of steelwork in buildings – Part 1: Code of practice for design – Rolled and welded sections.* London: BSI, 2001

5.6 Johansen, K.W. *Yield line theory.* London: C&CA, 1962

5.7 Alexander, S. 'New approach to disproportionate collapse', *The Structural Engineer*, 82(23), 7 December 2004, pp14-18

5.8 Corley, W.G. et al. 'The Oklahoma City bombing: summary and recommendations for multihazard mitigation'. *ASCE Journal of Performance of Constructed Facilities*, 12(3), August 1998, pp100-112

5.9 Corley, W.G. 'Effects of structural integrity on damage from the Oklahoma City, USA bombing'. In Neale, B.S. ed. *Forensic Engineering: the investigation of failures.* London: Thomas Telford, 2001, pp1-11

5.10 *BS EN 1990: 2002: Eurocode: Basis of structural design.* BSI, London 2002

5.11 *BS 8110-1: 1997: Structural use of concrete – Part 1: Code of practice for design and construction.* London: BSI, 1997

6 *In situ* concrete: issues and solutions

Key reference: *How to design concrete buildings to satisfy disproportionate collapse requirements* O Brooker, The Concrete Centre 2008[6.1]. BS EN 1992-1-1: Eurocode 2[6.2] (EC2) and BS 8110-1[6.3] are also key references.

6.1 Introduction

Well detailed, *in situ*, reinforced concrete is inherently robust[6.3, 6.4]. In most cases, checks for compliance with tying rules will show that tying requirements have already been met through the normal reinforcement provided. Therefore in a large part, this chapter discusses best practice in the positioning and detailing of tie reinforcement rather than how to provide for minimum code compliance.

In precast, and hybrid precast and *in situ* construction, joints between units often form a reinforcement path discontinuity. Where this occurs, the provision of a continuous tie will need more explicit consideration. Industry best practice methods of providing such continuity are presented in Chapter 7.

Occasionally it may be necessary to consider a concrete member as a key element or in other cases to consider the implications of element removal on the rest of the structure. The appropriate approach to these situations is discussed in Chapter 5 and below.

This chapter broadly follows the requirements of BS 8110[6.3]. However, for the UK, the rules in EC2[6.2] as supplemented by the UK National Annex[6.5] and PD 6687[6.6] require near identical consideration.

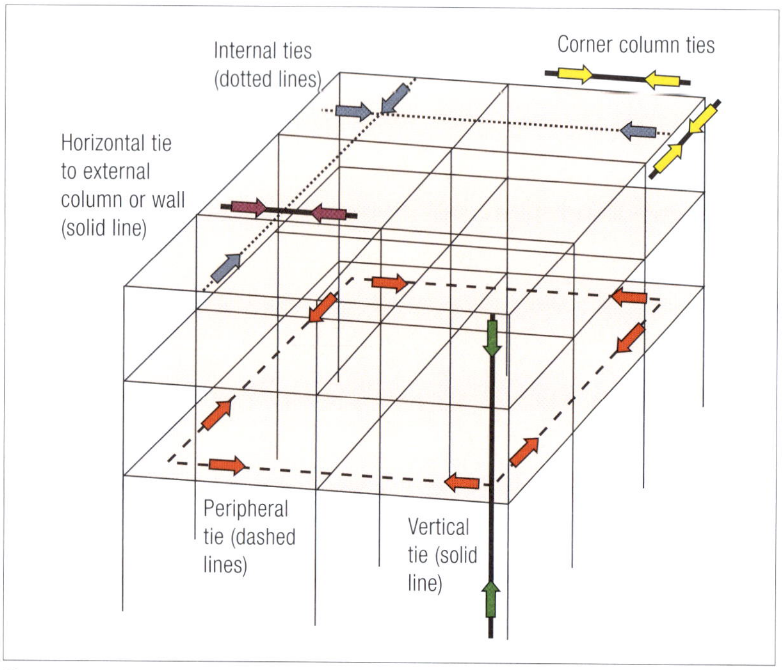

Figure 6.1 Ties required in concrete structures

Furthermore, whilst the values of specific parameters may vary, the approach described is good practice wherever construction takes place and should be considered a minimum requirement in the absence of any more onerous local regulatory requirements.

The rules incorporated in BS 8110[6.3] are semi-empirical. Whilst it may be useful to rationalise mechanisms to aid understanding, such rationalisation should not be used to reduce the stated requirements.

6.2 Overall robustness strategy

At the highest level, the robustness strategy advocated for concrete structures is similar to that for other materials as will be set out in later chapters. Likewise, the sequence of design for robustness is as set out in Chapter 5.

6.3 Notional horizontal load

In BS 8110[6.3], the notional horizontal loads are defined as a minimum of 1.5% of the characteristic dead load. This criterion can be significant for the design of low rise buildings and in the long direction of narrow, higher rise, buildings. The 1.5% effect is considered as a minimum load and ignored if the wind load is greater. In BS 5950[6.7] the corresponding load is 0.5% of the factored vertical dead and imposed loads, applied at the same level, thus the force value is expressed differently but is nevertheless numerically similar.

EC2[6.2] has a similar concept to BS 8110[6.3] except that the horizontal load is developed from a notional out-of-verticality (see Section 5.3).

6.4 Tying

6.4.1 Provision of ties

Four types of ties (as shown in Figure 6.1) are specified both in BS 8110[6.3] and EC2[6.2]; all are discussed below. Ties are normally assumed to act at their characteristic strength and no other actions are considered in conjunction with the tie force. This means that bars provided for other structural effects can be included in the area of reinforcement assumed for the tie. For design, the tie force magnitude can be derived from References 6.1, 6.2 or 6.3. The magnitude varies with the number of storeys but will not exceed 60kN for peripheral ties and perhaps double that for internal ties.

6.4.2 Peripheral ties

BS 8110[6.3] introduced the concept of peripheral ties. Arguably the edge of a structure is the most vulnerable to damage and moreover has reduced opportunities for developing alternative load paths via two way spanning. Hence provision of a peripheral tie ensures an alternative load system in the edge of the structure. The peripheral tie also provides a zone in which internal ties can be anchored and ensures that perimeter vertical elements are interconnected with the main tying system. Peripheral ties should also be provided around any large slab openings, such as those for atria. BS 8110[6.3] and EC2[6.2], as implemented in the UK, define the tie force magnitude and partial factors that can be used in evaluating tie capacity ($\gamma_s = 1$ for reinforcement and prestressing steel); the tie force value F_t is typically 60kN (ultimate) which is easily carried by two H10 bars. The resistant reinforcement should be located in the outer 1.2m of the slab or within perimeter beams or walls.

Peripheral tie value (derived from BS 8110):
$F_t = (20 + 4n_o) \leq 60$ (where n_o is the total number of storeys in the structure)

Note the units. The F_t value has no units. The force value is $1.0F_t$ in kN (60kN) which is to be located in the slab edge 1.2m.

6.4.3 Internal ties

Internal ties should be provided in two orthogonal directions. To maximise their benefit, ties should be as ductile as possible and ideally placed towards the bottom of the section as tests and experience have shown bottom bars to be more effective. Tie ductility can be improved by using higher ductility reinforcement, e.g. Class B or C. The maximum spacing of internal transverse ties is $1.5l_r$ where l_r is the greater spacing between columns. However, it is generally beneficial to adopt a lower spacing, indeed the requirement for these ties to interact with column ties (see Section 6.4.6) means that a practical maximum is the column spacing.

For internal ties (as distinct from peripheral ties), the tie force is evaluated differently and has different units. For internal ties, the force is derived in kN/m width (of the slab) whereas at the edge it is a defined force confined to a defined width. Hence:

F_{tie} is the greater of $(1/7.5) (g_k + q_k) (l_r/5) F_t$
or
$1.0F_t$ defined as kN/m across the internal slab width (this will not exceed 60 kN/m)

where:
$(g_k + q_k)$ is the sum of the average permanent and variable floor loads (in kN/m²). Note the variable load q_k is not in this case reduced (see Section 5.13)
l_r is the greater of the distances (in m) between centres of the columns, frames or walls supporting any two adjacent floor spans in the direction of the tie under consideration
$F_t = (20 + 4 n_o) \leq 60$ (where n_o is the total number of storeys in the structure).

6.4.4 Horizontal ties to external walls and columns

To address the possibility of walls or columns being pushed outwards, each external wall or column should be tied back into the main structure. In the case of a wall incorporating a peripheral tie, the requirement is to ensure that the internal ties are anchored into the peripheral tie. In all other cases, it is recommended that the column/wall ties are lapped with the internal ties. The tie minimum capacity is given in the appropriate code; these are either notional amounts or a percentage of the column/wall load. For corner columns, the tie force should be provided in two directions approximately at right angles to each other.

6.4.5 Vertical ties

As discussed in Chapter 5, vertical ties provide the opportunity, in the event of lower column removal, for floor loads to be supported by hanging from the column above; they also provide a minimum level of robustness to inhibit column removal. Vertical ties should be provided within every column and to each wall carrying vertical load. The tie force to be resisted is equal to the ultimate design load carried on any floor level by the element (calculated with the accidental load factors, see Section 5.13).

Where the vertical element is supported at its lowest level by anything other than a foundation, the overall robustness of the support should be considered and most likely the supporting element will be designed as a key element or perhaps an analysis will be carried out to demonstrate that removal of the supporting element does not lead to disproportionate collapse.

6.4.6 Continuity of ties

Ties should be effectively continuous. This means that where bars forming ties are lapped, the detailing should be such that failure is always in the bar and not in the lap thus ensuring a ductile performance. In practice, this means that laps should always be designed for the bar full capacity even if the required tie force is lower. Additionally, where bars forming the tie system are notionally lapped but not adjacent, the lap length should be increased and the need for links considered, as described in EC2 (Clause 8.7[6.2]). This is to ensure that an effective strut and tie system can be formed between the bars in the lap zone.

For tying systems to work, different types of ties must interact and the need for column (vertical) ties, internal and peripheral ties to be linked has been discussed above (Sections 6.4.1 to 6.4.5). It is essential for vertical and horizontal ties to interact if catenary action is to be developed. Such interaction can generally be presumed if some or all of the tie steel in each direction passes through the column, so it is recommended that horizontal ties are placed in the bottom of the slab or beam at the column location. It is also worth highlighting that if the vertical tie is to share load up the building, the connection to the higher floors needs to be capable of taking reverse shear. In other words, if a floor is hung from the column above, the column is pulling down on the floor above, i.e. the shear is reversed from the normal situation where the column would be pushing up on the floor. This can complicate some hybrid type connections (Chapter 7 incorporates a number of details).

6.4.7   Additional considerations for post tensioned concrete

The principles for post tensioned construction are identical to those adopted for reinforced concrete. In bonded prestressed concrete, continuous tendons provide an excellent tie since there are no or fewer laps. For bonded construction, the principal challenge is to ensure sufficient interaction between the horizontal and vertical ties. Research[6.8] has shown that a significant improvement in post failure capacity occurs when tendons pass directly over the column; however this is not always possible. In this situation, it is recommended that ducts are placed as closely as possible (either side of the column), and additional bottom steel is provided through the column in each direction to lap onto the duct line. The tendons support at midspan as they are in the bottom there.

For unbonded tendons, a failure of tendons in one bay may lead to failure in adjacent bays. For this reason, it is not appropriate to consider unbonded tendons as part of the tying system, and so tying should be provided wholly with normal reinforcement.

6.5   Element removal and key element design

An alternative to following the prescriptive rules for tying is to consider the removal of an element and to investigate the subsequent collapse area following the principles of Chapter 5. There are no special requirements for *in situ* concrete structures excepting that partial factors for materials can be reduced.

In EC2[6.2], for the accidental load case, the partial factors are as follows:

γ_c for concrete = 1.2 and
γ_s for steel = 1.0.

In BS 8110[6.3] the γ_c for concrete under exceptional loads is only reduced for flexure (= 1.3), as the partial factor for shear is lower anyway. BS 8110 adopts a partial factor γ_s for steel = 1.0, as does EC2.

There are no special requirements for concrete key element design and the approaches described in Section 5.11 should be adopted.

6.6   Good detailing practice

There is an implicit assumption of structural ductile performance for the nominal Approved Document A[6.9] tying recommendations to work in practice, not least since catenary action presumes significant axial and rotational ductility. To some extent, codes recognise the demands of ductility since a modest amount of moment redistribution is permitted in continuous beams and since slabs designed on yield line methodology presume rotation at hinge positions. In routine design, ductility demands are not calculated explicitly, rather they are catered for by assuring members are under-reinforced and by

application of detailing rules. In cases where ductile energy absorbing capacity is required for extreme robustness (as in resistance to blast or earthquakes) advanced rules[6.10, 6.11] are available.

All relevant concrete codes include requirements for minimum reinforcement. In the cases of direct tension and flexure, these minimum reinforcement contents ensure that section capacities after concrete cracking are approximately equal to or greater than those which existed before cracking. Whilst this is a prerequisite to controlling crack widths, minimum reinforcement also ensures that there is a reasonable amount of post-cracking ductility within the element. In the event of overload, this ductility provides warning and facilitates load shed to other elements preventing gross failure. Similarly, design codes require minimum amounts of reinforcement at supports; even where no moment has been assumed. These requirements, again driven by the need to control cracking, also provide alternative (or enhanced) load paths. Such additional reinforcement provides protective strength against the possibility of moment reversal which can be a feature in accidental loading[6.12].

Minimum percentage of reinforcement in columns assures a minimal tensile capacity for what are supposedly compression members. Minimum link requirements in columns are defined to give a certain level of ductility at the column/floor connection. In extreme cases, as in seismic design, greater ductility is demanded and a greater number of links are provided and better link anchorage is specified. Under overload, such links prevent column bars buckling and confine the concrete core, preventing disintegration, so adding to the section's rotational capacity whilst maintaining moment resistance. Nonetheless, even the non-seismic link provision will ensure an amount of column rotation is possible prior to failure.

It can be seen from above that there are a number of detailing requirements in current codes, related to minimum reinforcement areas which add to the inherent robustness of concrete buildings. It is important that where such minimum steel requirements are not provided, say due to new construction techniques or systems, the effect on the overall structural robustness be re-considered.

6.7   References

6.1   Brooker, O. *How to design concrete buildings to satisfy disproportionate collapse requirements.* Camberley: The Concrete Centre, 2008

6.2   *BS EN 1992-1-1: 2004: Eurocode 2: Design of concrete structures – Part 1-1: General rules and rules for buildings.* London: BSI, 2004

6.3   *BS 8110-1: 1997: Structural use of concrete – Part 1: Code of practice for design and construction.* London: BSI, 1997

6.4   Institution of Structural Engineers. *Standard method of detailing structural concrete: a manual for best practice.* 3rd ed. London: Institution of Structural Engineers, 2006

6.5   *NA to BS EN 1992-1-1: 2004: UK National Annex to Eurocode 2: Design of concrete structures – Part 1-1: General rules and rules for buildings.* London: BSI, 2005

6.6   *PD 6687: 2006: Background paper to UK National Annex to BS 1992-1.* London: BSI, 2006

6.7   *BS 5950-1: 2000: Structural use of steelwork in buildings – Part 1: Code of practice for design – Rolled and welded sections.* London: BSI, 2001

6.8   Pinho Ramos, A. and Lucio, V.J.G. 'Post punching behaviour of prestressed concrete flat slabs'. *Magazine of Concrete Research*, 60(4), May 2008, pp245-251

6.9   Office of the Deputy Prime Minister. *The Building Regulations 2000. Approved Document A: Structure.* London: NBS, 2004

6.10 Park, R. and Paulay, T. *Reinforced concrete structures.* New York: Wiley-Interscience, 1975

6.11 *ACI 318-08: Building code requirements for structural concrete (ACI 318-08) and commentary.* Farmington Hills, MI: ACI, 2008

6.12 Beeby, A.W. 'Safety of structures and a new approach to robustness'. *The Structural Engineer*, 77(4), 16 February 1999, pp16-21

7 Precast concrete: issues and solutions

7.1   Introduction

Structures formed of loadbearing precast panels will be forever associated with the Ronan Point collapse[7.3]. Although that failure is historically interesting, it is entirely relevant to observe that nowadays ample opportunities exist for precast members to be interlinked and tied together, so as to minimise the risk of future disproportionate collapse; this is particularly so where precast units are subsequently concreted in to develop composite action. In the years since Ronan Point, many effective tying and restraint details have evolved and some typical details are given in this chapter in Section 7.3.

Nevertheless, precast construction lacks the automatic continuity inherent with *in situ* construction, so the provision of robustness requires more direct engineering intervention. Particular care is required during the construction phase, partly because the final *in situ* additions will be absent and partly because many typical support systems can be torsionally unbalanced, for example if precast units are added to one beam side only, and designers need to consider instability issues as part of their obligations under CDM2007[7.4] (see Chapter 3).

The principles and strategies to be deployed for preventing collapse are those set out in previous chapters, especially Chapter 5. The relevant codes (BS EN 1991-1-2[7.5] and BS EN 1991-1-7[7.6] and BS 8110[7.2]) for precast concrete provide design rules that have to be followed.

7.2   Class 1 and Class 2A and 2B buildings

7.2.1   Class 1 and 2A buildings

The requirements for Class 1 and 2A buildings are fundamentally different from those required for Class 2B (and Class 3 buildings). The Building Regulations[7.7] (Part A3) require that Class 2A buildings incorporate effective anchorage of the floors *or* effective ties (there is no such prescription for Class 1 but lack of anchorage would lead to instability in walls running parallel to floor spans). Clause 5.1.8.3 of BS 8110[7.2] requires that the dead weight of members is effectively anchored to that part of the structure containing the ties. This design force should be based upon the accidental load case

reaction i.e. $1.05(1.00G_k + 0.33Q_k)$. Most designers opt for providing effective floor anchorage to the walls/beams rather than specifying ties. Such anchorage can be achieved in most cases simply by the friction between the floor and the wall/beam although consideration of parameters such as temperature and camber may negate this approach. Where a floor unit is prestressed, its camber prevents proper contact with walls below running parallel to the span and so friction can only be relied on when bedding mortar is used.

7.2.2   Class 2B buildings

The provision of effective horizontal and vertical ties is the generally preferred solution. The fully tied solution is based on the assumption that precast structures will get sufficient robustness to withstand a moderate degree of abnormal loading through the standard tie network. The ties required are those described in Chapter 5 and the load path for all ties must follow the requirements set out in Chapter 6.

Tying can generally be accomplished by utilising the members themselves with appropriate end connections. Beams designed to carry floor or roof loading will normally be suitable as continuous ties. For example, a series of beams in line will act as an internal tie provided end connections to intermediate elements (beams or columns) have the required strength. Alternatively, ties may be added as steel members or as steel reinforcement embedded in concrete strips set between precast units or set within recesses purposely left in the ends of the precast units (see details in Section 7.3). Crucially, reinforcement provided for other purposes may be legitimately utilised for ties; it may be unnecessary to add additional steel.

Precast floor units are heavy yet could still be dislodged if their supports move in an explosion or during some other partial collapse. To prevent this, BS 8110[7.2] Clause 5.1.8.3 explicitly requires that units must be tied, either to each other over a support, or to the support itself. Normally this is achieved through provision of an *in situ* concrete topping with added mesh, or via looped reinforcement at the slab ends (acting to form an emergency hinge or bearing) and subsequently encased. The ability of the shear interface to tie the precast to the *in situ* topping must be checked. Loops should be preferably in the form of securely anchored hairpins placed in the middle of the floor depth to allow for maximum efficiency and deformability (reinforcement encased within a top screed is unsuitable for utilisation in catenary action). Typical details are shown in Section 7.3.

Where *in situ* topping is not provided, reinforcement can be provided within plank end pockets with bars then grouted up to link units together.

Precast stairs are referenced in BS 8110[7.2] Clause 5.1.8.3: "where precast floor, stair, or roof members" and in PD 6687[7.8]: "precast floor and roof units and stair members". Precast stairs are the primary means of escape: they are clearly vulnerable and should be treated as other floor units. Since stairs are often added early to provide access routes, care is required over their temporary stability including the effect of tolerances.

Precast units invariably span one way and shed no load onto walls in line with their span. Thus it is unsafe to assume that any walls below will be restrained by friction against out-of-plane movement unless bedding mortar/positive anchorage is provided.

7.3   Typical details

7.3.1   General

Typical details are available for the use of precast slabs in masonry, steel and concrete buildings for Class 2A and 2B buildings. These are shown in Figures 7.1 to 7.10 inclusive. The principles of the details are compliant with both EC2[7.1] and BS 8110[7.2].

7.3.2   Precast floor/masonry walls Class 2A buildings

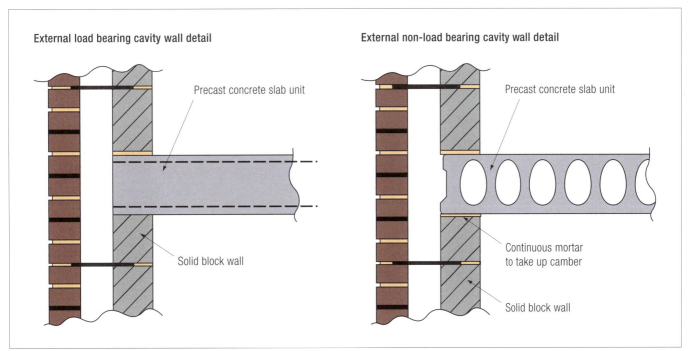

Figure 7.1   Provision of effective external wall anchorage (via embedment). Class 1 and 2A

7.3.3   Precast floor/masonry walls Class 2B buildings

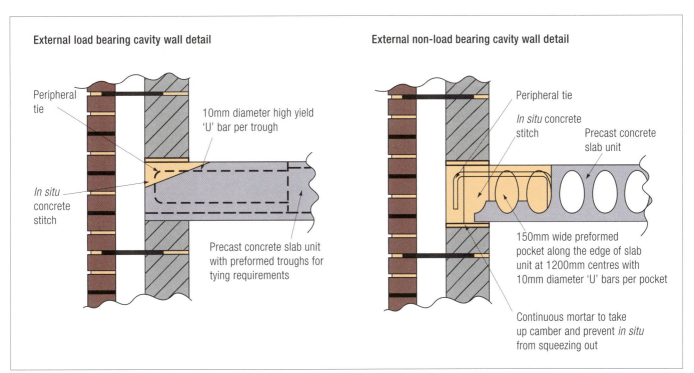

Figure 7.2   Provision of horizontal tying to external walls, Class 2B

7.3.4   Precast floors/steel framed Class 1 and 2A buildings and 2B buildings

BS 5950-1:2000[7.9] requires in Clause 2.4.5.3 (e) that:

> "where precast concrete or other heavy floor, stair, or roof units are used they should be effectively anchored in the direction of their span, either to each other over a support, or directly to their supports as recommended in BS 8110".

Typical details are shown in Chapter 8.

7.3.5   Precast floors/concrete framed Class 2A buildings

Figure 7.3 shows effective ties from the floors to the walls.

7.3.6   Precast floors/concrete framed Class 2B buildings

The cross wall system (see Figure 7.4) illustrates typical connection requirements (further details can be obtained from References 7.10 and 7.11). Vertical ties are provided via vertical bars cast into the walls which pass through the floor area. The wall units above the floor can be designed to span horizontally in the event of a loss of wall support below. Reference 7.10 contains these details.

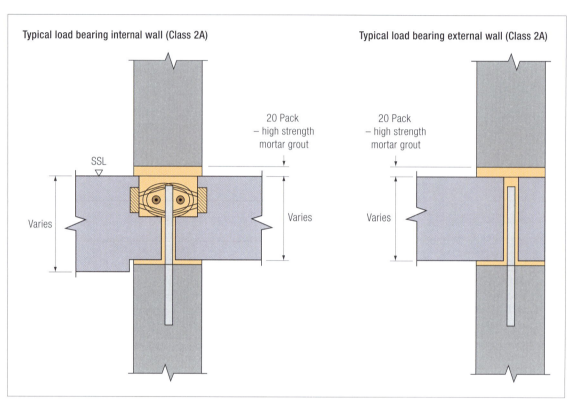

Figure 7.3   Ties from floors to walls, Class 2A

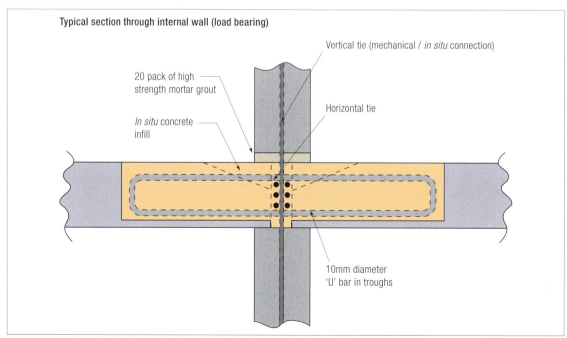

Figure 7.4   Ties from floors to walls, Class 2B

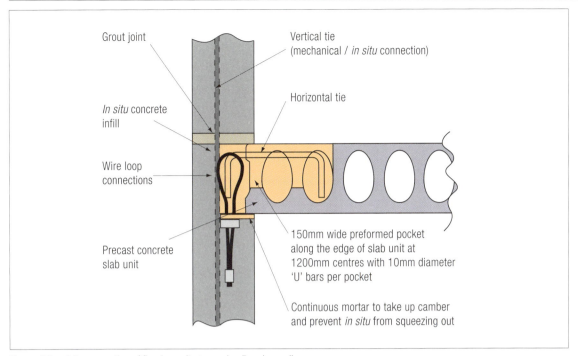

Grout joint

Vertical tie
(mechanical / *in situ* connection)

In situ concrete
infill

Horizontal tie

Wire loop
connections

150mm wide preformed pocket
along the edge of slab unit at
1200mm centres with 10mm diameter
'U' bars per pocket

Precast concrete
slab unit

Continuous mortar to take up camber
and prevent *in situ* from squeezing out

Figure 7.5   Interconnection of flooring units to non loadbearing walls

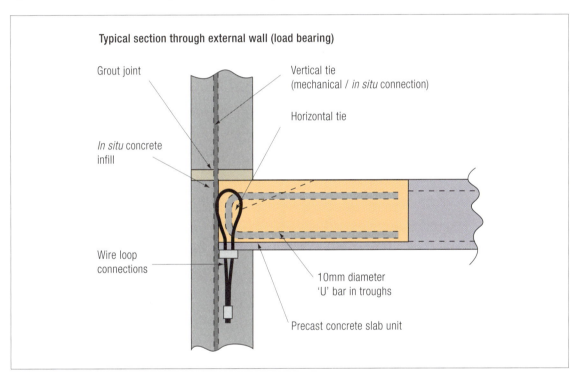

Typical section through external wall (load bearing)

Grout joint

Vertical tie
(mechanical / *in situ* connection)

In situ concrete
infill

Horizontal tie

Wire loop
connections

10mm diameter
'U' bar in troughs

Precast concrete slab unit

Figure 7.6   Interconnection of flooring units to loadbearing walls

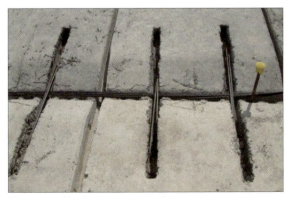

Figure 7.7   Ties between precast units

Figure 7.8   Interconnection of flooring units to walls via site
connection (side pocket in hollow core unit), wall parallel to
floor span

7.3.7   Typical tie details: beam and column frame

Beam and column frame solutions require careful detailing to manage the effects of torsion and excessive movements particularly during construction and before any permanent continuity steel is effective (see Figures 7.9 and 7.10). Torsion and movement can be induced for example when there is a lack of support to the beam toe during construction. Movement can also be induced by temperature on exposed slabs. Such movements can cause spalling, loss of bearing and cracking in the topping. Shear can be critical when the slab support beam rotates and induces an additional transverse shear in the floor slab. The existence of effects like these emphasise the need for robustness in the detailing.

Details (shown in Figures 7.9 and 7.10) tie the hollow core floors into the beam and, when fully concreted, help with the torsion and restrain the beams from moving. This detail requires that the beam is torsionally stable during construction. Certain proprietary details have been tested to demonstrate their strength (see Figure 7.11 where the end connection shown has been tested up to 900kN ultimate load).

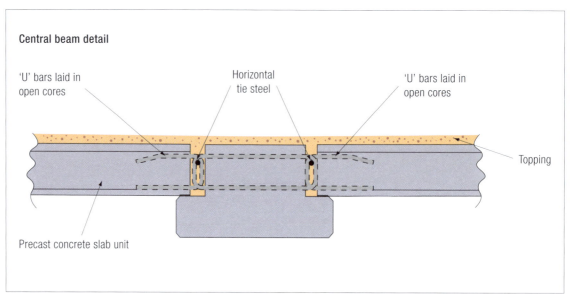

Figure 7.9   Interconnection between precast units onto an internal support beam (Class 2B beam and column frame typical details)

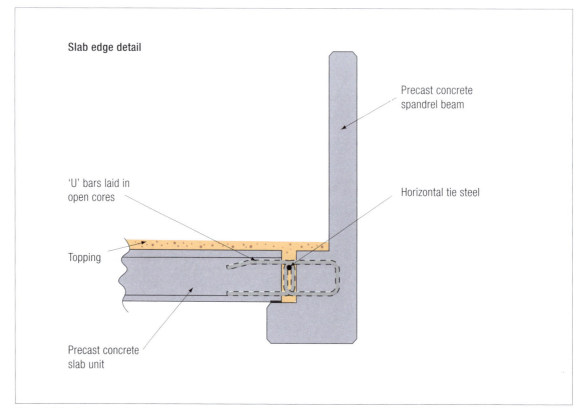

Figure 7.10   Interconnection between precast units onto an edge beam (Class 2B Beam and column frame typical details)

| Photograph of beam prior to installation | Photograph of beam with floor installed |

Figure 7.11 Proprietary beam end fixings before and after installation

7.4   Class 2B buildings: notional removal of elements

Following the general guidelines of Chapter 5, Class 2B buildings may be assessed by the notional removal of elements. A notional removal concept is a fairly natural one in precast assemblies especially where some parts may be non-loadbearing. The notional removal route will require increased attention to detail but in compensation, many panel structures are able to function as deep beams and so provide good spanning capability.

Where vertical ties have been incorporated, a support system is possible via the suspension of any newly unsupported elements back to intact structure above the damaged area. Alternatively, support may be achieved via catenary or cantilever action of the surrounding structure; this is particularly useful in the case of corner column failure where horizontal tie reinforcement added on top of composite floor beams can take up the cantilever tensile stresses.

To justify these survival systems, the assumed tensile reinforcement must be end-anchored, for example via inside projecting U bars within the top of the unit. The anchorage of ties used in catenaries should be demonstrated (to the relevant code) and not just assumed.

A standard support possibility is to assume bridging of damaged areas by catenary action as described in Chapter 5. Loss of a central support will typically mean that beams effectively double in span but the consequent excess forces may still be carried, albeit with increased deformation. For precast units, the critical engineering aspects are to assure end anchorage (which implies continuity of reinforcement) and to prevent longitudinal top reinforcement bursting upwards at the point of maximum sag. To avoid this, links are required as shown in Figure 7.12 and the column immediately above the lost support has to be checked for interconnection and tension. The design objective is that stitching the various components together will make the frame behave more like an *in situ* frame.

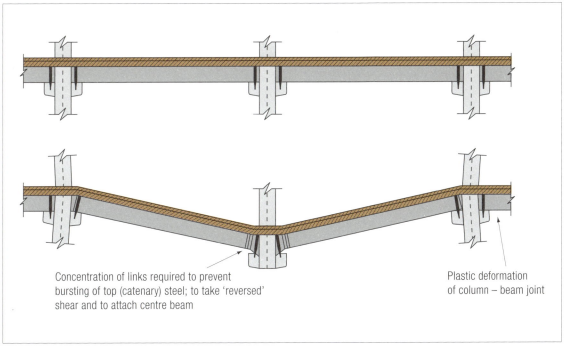

Concentration of links required to prevent bursting of top (catenary) steel; to take 'reversed' shear and to attach centre beam

Plastic deformation of column – beam joint

Figure 7.12 Catenary action in precast floor beams (with composite action)

Under similar loss of support scenarios, prestressed beams spanning onto corbels would probably keep their original shape without significant cracking and would just separate from any corbel above the destroyed column unless anchored down. Rotations/deformations in the column joint zone would probably be large, so the assumptions on survival are only valid if the anchorage details are sufficiently ductile. Manufacturer's propriety connections must take this into account.

Wall frame mechanisms can provide for alternative load paths i.e. the same survival systems used for skeletal structures also function in preventing the collapse of loadbearing wall structures. The survival capability of such structures is often high because of the large cantilevering and bridging capability of wall panels, effectively acting as deep beams. The following support systems can operate (as illustrated in Figure 7.13):
- Suspension of the elements from the intact upper structure above the damaged area. This is validated by vertical ties from foundation to roof level in all walls.
- Cantilever action of the surrounding structure. For example, in the case of corner wall panel failure, the horizontal tie reinforcement on top of the wall panel will take up the tensile stresses of the cantilever. To be effective, the tie-reinforcement must be connected into the wall panel, for example inside hairpins projecting above the unit tops.
- Bridging of the damaged area by the intact wall panels above.

7.5   Key elements

The standard approach for key elements set out in Chapter 5 can be adopted for precast units and that design approach is required wherever local collapse exceeds permitted area limits.

7.6   References

7.1   *BS EN 1992-1-1: 2004: Eurocode 2: Design of concrete structures – Part 1-1: General rules and rules for buildings*. London: BSI, 2004

7.2   *BS 8110-1: 1997: Structural use of concrete – Part 1: Code of practice for design and construction*. London: BSI, 1997

7.3   Ministry of Housing and Local Government. *Report of the enquiry into the collapse of flats at Ronan Point, Canning Town*. London: HMSO, 1968

7.4   *The Construction (Design and Management) Regulations*. Norwich: The Stationery Office, 2007 (SI 2007/320)

7.5   *BS EN 1991-1-2: 2002: Eurocode 1: Actions on structures – Part 1-2: General actions – Actions on structures exposed to fire*. London: BSI, 2002 and *NA to BS EN 1991-1-2:2002: UK National Annex to Eurocode 1: Actions on structures – Part 1-2: General actions - Actions on structures exposed to fire*. London: BSI, 2007

7.6   *BS EN 1991-1-7: 2006: Eurocode 1: Actions on structures – Part 1-7: General actions – Accidental actions*. London: BSI, 2006 and *NA to BS EN 1991-1-7: 2006: UK National Annex to Eurocode 1 - Actions on structures – Part 1-7: Accidental actions*. London: BSI, 2008

7.7   *The Building Regulations 2000*. London: The Stationery Office, 2000 (SI 2000/2531), as amended by *The Building (Amendment) Regulations 2004*. [s.l.]: The Stationery Office, 2004 (SI 2004/1465)

7.8   *PD 6687: 2006: Background paper to UK National Annex to BS 1992-1*. London: BSI, 2006

7.9   *BS 5950-1: 2000: Structural use of steelwork in buildings – Part 1: Code of practice for design – Rolled and welded sections*. London: BSI, 2001

7.10   Brooker, O. and Hennessy, R. *Residential cellular concrete buildings: a guide for the design and specification of concrete buildings using tunnel form, crosswall or twinwall systems*. Camberley: The Concrete Centre, 2008

7.11   Whittle, R. and Taylor, H. *Design of hybrid concrete buildings: a guide to the design of buildings combining in-situ and precast concrete*. Camberley: The Concrete Centre, 2009

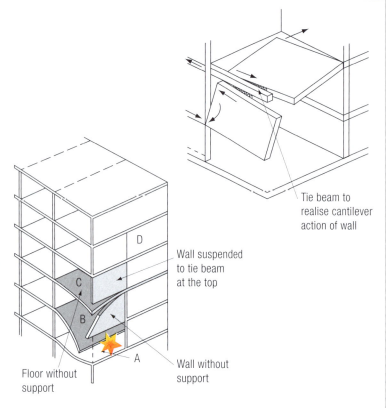

Tie beam to realise cantilever action of wall

Wall suspended to tie beam at the top

D

C

B

A

Wall without support

Floor without support

Notes

A = Floor with support but possible impact from debris
B = Floor displaced due to lack of support
C = Floor without support from below but tied to wall
D = Wall cantilevering to provide support to frames above

Figure 7.13   Alternative means of protection against progressive collapse in wall frame structures

8 Steel: issues and solutions

Key reference: SCI Publication P341 *Guidance on meeting the robustness requirements in Approved Document A (2004 Edition)* [8.1]. BS EN 1993-1-1: Eurocode 3 [8.2] (EC3) and BS 5950-1 [8.3] are also key references.

8.1 Introduction

Normal, well detailed, steel frames should be inherently robust benefiting from being tied together in a similar manner to concrete frames and this tying offers opportunities for alternative load path development. Moreover, steel members and their connections should automatically possess a degree of ductility enabling frames to distort and absorb energy under unusual conditions. However, where connections are pinned rather than fixed there is less opportunity for the redistribution of moments and caution is required as the attributes of strength and ductility are dependent on the frames having appropriate connections. Appropriate connections possess characteristics of strength and are detailed so that they have the ability to deform whilst still retaining load carrying capacity. Unlike concrete structures, stability is more of an issue for structural steelwork and a sound grasp of what systems stabilise members, sub frames and whole structures is essential if overall robustness is to be guaranteed.

It is of course essential that the structural form adopted provides consistent load paths to ground for vertical and horizontal forces.

In pin jointed frames reliant on bracing or propping for global stability, some redundancy is required in the stability system or a judgment made that the overall prop (say a concrete core or a bi-steel core) is strong enough to function as a single robust point (see Figure 8.1).

Traditional steel design approaches have in the past assured a measure of robustness; thus clauses imposing minimum connection sizes, minimum weld sizes matched to material thickness, minimum slenderness ratios and so on all helped considerably to add robustness to frames. However, current codes (both BS 5950 [8.3] and EC3 [8.2]) have tended to delete clauses imposing minimum slenderness ratios. This consequently imposes more obligations on the designer to consider the robustness of the system as a separate issue having regard to the circumstances of use. Even the traditional method of sizing lateral restraints based on a percentage of compression force has its origins not in precise science, but as a rule of thumb to assure a sensible system having adequate strength and stiffness. It is not sound engineering to prejudice the stability of a steel frame by skimping on restraint when the provision of substantial restraint can be gained for little cost.

Figure 8.1   Modern steel frame stabilised by a single bi-steel core

Such traditional concepts should not be discarded, especially in regards to connection capacity. It is sound engineering to design connections to minimum force levels, to keep connections in proportion to the members they join, irrespective of calculated demand, and not to use welds that are too small and sensitive to minor variations in throat size. Connection detailing should be such that ductility is in-built so that inherent assumptions about load re-distribution can be realised. When considering resistance against collapse, the possibility of moment/shear reversal on joints has to be borne in mind. Having noted that, industry standard connections will comply with robustness objectives for normal buildings. The ability of connections to carry load under conditions of severe distortion may need to be examined explicitly in unusual structures.

Once complete, steel frames ought to be inherently robust, especially when they are of composite construction or can benefit from the diaphragm action of the floors and even cladding (acting as stressed skin or purlin restraints) when this is well fixed. Probably the most vulnerable stage of the steel structure's life is the erection phase and there have been cases of progressive and disproportionate collapses simply because global stability was inadequately provided for during construction. It might even be argued that the infamous box girder bridge collapses of the 1960s exemplified lack of robustness, since failure of a relatively minor detail was capable of precipitating a gross change in state. Codified rules are not intended to address robustness issues during construction so the construction design teams need to address the issue directly.

To assure robustness during construction and in-service, the design engineer should be clear on what provides stability and should ensure that the load path for stability is continuous, back to a robust foundation as a pre-requisite. Thereafter, the specific checks for robustness need to be carried out.

There are no specific requirements in EC3[8.2] for robustness so the requirements are those derived from EN 1990[8.4] and Eurocode 1[8.5] and BS EN 1993 parts 1[8.2] to 8[8.6]. The main differences to current UK practice are:
– The horizontal ties required for Class 2A buildings are the same as those for Class 2B buildings.
– There is no reduction in horizontal tie forces for buildings less than 5 storeys.
– The tie force is based on a similar expression to BS 5950[8.3] using the uniform load but it uses the accidental load case instead. The coefficients used are different to BS 5950 but in general, the end values are similar though dependant on the values of permanent and imposed load.
– There is no 'deemed to satisfy' rule about making the tie force equal to the shear force.
– The vertical tie force required is the maximum load from any floor attached to the column, not just that between splices.
– The vertical tie force uses the accidental load case so will generally be less than in BS 5950.
– There are no particular requirements for redundancy in stability systems.
– The requirements for tying precast floor, roof and stair units are in PD 6687:2006[8.7].

(The Eurocode uses a different formula for the notional horizontal forces which, with all the modification factors included, tends to give a smaller force than those obtained from BS 5950[8.3]. However the EC3[8.2] value is used in all load combinations which makes sense because the lack of column verticality is related to steel erection and not the load combination).

The key reference[8.1] describes in detail how the regulations[8.8] are to be interpreted for their application to steel structures of every class and the reference gives worked examples to foster understanding. However, Reference 8.1 only deals with hot rolled sections; SCI Advisory Note AD280[8.9] provides guidance on achieving structural integrity for light gauge steel structures designed to BS 5950 Part 5[8.10] and Reference 8.11 gives guidance on Slimdek members.

BS 5950[8.3] contains specific clauses aimed at assuring robustness plus compliance with UK regulations[8.8]. It is a specific aim of the design (Clause 2.1.1.1[8.3]) that: "the structure should be designed to behave as one three dimensional entity" and "the layout ….should constitute a robust and stable structure under normal loading to ensure that, in the event of misuse or accident, damage will not be disproportionate to the cause." Clause 2.1.1.2[8.3] relates to overall stability and emphasises the need to identify the designer with overall responsibility.

As with all other codes, the provisions for robustness are largely empirical. Reference 8.1 stresses the use of informed engineering judgement rather than promoting compliance by close attention to rules.

Particular clauses of relevance in BS 5950[8.3] are:
– Clause 2.4.2.3 *Resistance to horizontal forces*
– Clause 2.4.2.4 *Notional horizontal forces*
– Clause 2.4.5 *Structural integrity* (along with sub clauses 2.4.5.1 to 2.4.5.4)

BS 5950[8.3] Clause 2.4.2.3 along with Clause 2.4.2.4 specifies a minimum value of horizontal load that has to be resisted. In the absence of wind (Combination 1) this is 0.5% of the factored vertical dead and imposed loads applied at the same level and this is deemed to account for practical imperfections likely to cause sway (such as lack of verticality). Where there is a wind load (Combinations 2 and 3), the horizontal component of the factored wind force should exceed 1% of the factored dead load applied at the same level. Clause 2.4.2.4 schedules situations where the horizontal forces are to be ignored and some structures for which the notional horizontal forces (NHF) have to be increased.

It can be shown that the notional horizontal forces (0.5%) equate to a column plumb of 1 in 200 which is greater than the verticality tolerance customarily specified, thus the forces provide margins against other effects as well.

The approaches to building in robustness and avoiding disproportionate collapse for steel structures follow the same general principles as for all other structural materials. Section 2.4.5 of BS 5950[8.3] sets out the means by which steel structures may be designed following the three approaches:
– tying
– notional element removal
– designing key elements.

The approaches may be mixed and frequently columns can be proven as key elements since they have the inherent strength to survive.

Clause 2.4.5.1[8.3] defines that "it may be assumed that substantial permanent deformation of members and their connections is acceptable"; the objective is to survive. Thus it may be assumed that portions of the structure still standing may be yielded and unserviceable.

8.2 Classification of steel structures

Generally the definitions of Categories 1, 2A and 2B follow the descriptions in Chapter 4 of this *Guide*. Reference 8.1 provides additional guidance for steel structures and for example gives guidance on how to classify mezzanine floors.

8.3 Class 1 buildings

There will be few steel buildings that fall under Class 1 and the accepted norm is that provided such structures are designed to the applicable code for applied loading no further action is required. However, Reference 8.1 points out the words in the guidance of AD-A[8.8] that "no additional measures are likely to be necessary", so leaving open the possibility that they might be. This is a reminder that in designing to avoid disproportionate collapse, each structure should be considered on its own merits and informed engineering judgment must be used rather than blindly following set rules (as outlined in Chapter 2). In reality BS 5950[8.3] treats Class 1 the same as Class 2A and requires ties (and as described in Clause 8.1.8.5, in the Eurocode, the horizontal ties required for Class 2A buildings are the same as those for Class 2B buildings).

BS 5950[8.3] Clause 2.4.5.2 *Minimum requirements*, controls Class 1 buildings and this recommends that:
– All buildings should be effectively tied together at each principal floor level.
– Columns should be tied in two directions approximately at right angles to each other, at each principal floor level.
– All ties (arranged as close as practical along the edges of the building and along each column line) and their end connections, should be capable of resisting a factored tensile load of at least 75 kN. This applies to all horizontal members.
– Horizontal ties should also be provided at roof level, except where steelwork only supports cladding that weighs not more than 0.7 kN/m² and that carries only imposed loads and wind loads.
– Precast floor/roof units should have bearing details conforming with BS 8110[8.12] (typical details have been illustrated in Chapter 7, other details are given in Reference 8.1).

For a fuller list of requirements, consult Clause 2.4.5.2[8.3]. Clearly, although demanded by the code, these are sound practical rules for achieving robustness in any steel structure (and not just Class 1). The tie value of 75kN is easy to achieve especially if the traditional approach is followed of using at least two M20 bolts in every connection.

8.4 Class 2A buildings

Class 2A buildings cover all buildings not exceeding 2 storeys to which members of the public are admitted and which contain floor areas not exceeding 2000m² at each storey (see Chapter 4); when the floor area increases to 5000m², the building is reclassified up to 2B. Thus Classes 2A (and 2B) cover a vast number of low rise traditional portal frame sheds. In the main portal direction, frames which have been detailed to exhibit plastic behaviour capability under overload must be inherently robust. Such inherent benefits should not be squandered by ignoring the essential stability provision required for compression flange stability (particularly on the underside of haunches) or under providing for it. Nor should the structure as a whole be rendered prone to frame toppling by the provision of bracing panels lacking in strength or stiffness. Although tension-only systems can be used, they can lack the robustness and stiffness offered by tension/compression systems.

In portal frames (and frames generally) the column holding down bolts (and foundations) play an important role in providing robustness during construction and providing stability against collapse in case of fire. Foundations for portal frames must be robust against lateral movement and all foundations must be big enough to provide stability for single columns cantilevering upwards on first erection (especially where multiple storey height columns are planned). In multi storey construction, emergency framing action may be called on (e.g. when confronted by high winds). Hence, to provide for this contingency, all columns should always be anchored down to their foundations.

The primary approach to avoiding disproportionate collapse in steel framed buildings is to provide horizontal tying of the frame elements and in BS 5950[8.3], the tying requirements are the same for Class 1 and Class 2A (see Section 8.3); Clause 2.4.5.2 governs.

There is no formal requirement in BS 5950[8.3] to anchor floor units down for Class 2A (although bearing details have to comply with BS 8110[8.12]) but fixing decking units down provides added diaphragm action and Reference 8.1 suggests fixing guidance (see also details in Chapter 7). Steel decking can function by stressed skin action if adequately fixed and that provides an advantage under say the destabilising loads of wet concrete during construction.

If effective ties cannot be provided, assessment under the options for Class 2B is possible i.e. by using notional removal or key element design (but note the recommendations of the Task Group (in Chapter 4) that provision of ties is always required in Class 2B buildings unless there is good evidence to the contrary).

8.5   Class 2B buildings

The formal robustness requirements for Class 2B buildings are described in Chapter 4. The design strategies are the standard ones of providing notional horizontal and vertical tying to minimum standards, with the alternatives of considering notional member removal or designing for key elements. In BS 5950[8.3], Clause 2.4.5.2 *Minimum requirements* is applicable plus the demands of Clause 2.4.5.3 which specifically covers 'Limiting the effects of accidental removal of supports'. Clause 2.4.5.3[8.3] specifies five conditions which are:
– general tying
– tying of edge columns
– continuity of columns
– resistance to horizontal forces
– heavy floor units.

The code should be referenced for detail but if any of the first three conditions are not met, the robustness check reverts to one of notional member removal. The acceptance criterion is the standard one of assuring that removal of any single supporting member will not cause an unreasonable area of the structure to fall down. The rest of the building is only required to remain stable and not necessarily serviceable. If the removal of any supporting member would cause disproportionate collapse then it should be designed as a key element.

If design of key elements has to be invoked, then that is covered in BS 5950-1[8.3] Clause 2.4.5.4.

8.6   General tying

8.6.1   General

In Class 1 and Class 2A buildings, ties are required to be concentrated along the column lines. But in Class 2B buildings, the ties should be distributed over the floor width as well. Nevertheless, this need not be done provided that the beams on the column lines (and their connections) are designed for their additional share of panel tie forces which are derived from the tying formula in BS 5950[8.3] (which includes a parameter for beam spacing). However, if the total tying force is concentrated on the beam line, it tends to be too onerous for standard beam column connections and it is then more efficient to take benefit from the tie capability that exists on any beams lying between columns.

BS 5950[8.3] Clause 2.4.5.3 (a) *General tying*, describes which horizontal members should be designed as ties and defines the tensile loads that such ties and their end connections should be capable of resisting and this value has to be at least 75kN. The expression for the load in the ties is based on a uniform floor load. Where there are significant other types of load (e.g. façades) the advice is that these should be taken into account by adding an equivalent uniform load. BS 5950[8.3] also allows a simplification whereby if the steel members and their end connections are capable of resisting a tensile force equal to the end reaction under factored loads, then they can

be assumed to provide adequate tying. Where the number of storeys is less than five, the tie force is reduced and a table of reduction factors is given in Clause 2.4.5.3.

Tying capacities need not be provided entirely by the steel frame. For example, in composite construction, a certain amount of the required horizontal tying can be provided by the concrete slab reinforcement, if it has been designed and detailed for that purpose. SCI publication P213[8.13] provides guidance on utilising slab reinforcement within the connection design.

The structure at the ends of the ties does not need to be designed for the tie force (although this looks illogical, it is just part of the empirical approach). At edge beams and columns, the connection of the tie is designed to take the force but the edge beam or column itself is not. Nevertheless, to be effective, the definition of connection should include the local area of the member which is being connected. Some sensible care is required in proportioning the connections so as not to negate reasonable assumptions about pin ended performance.

8.6.2   Edge columns

Tying to edge columns is required to ensure that such columns remain attached to the building. BS 5950[8.3] Clause 2.4.5.3 (b) *Tying of edge columns*, states that ties connected to edge columns should be capable of resisting the larger of the following forces: the design loads for general tying specified in Clause 2.4.5.3 (a) or 1% of the factored vertical dead and imposed load in the column at that level (but this value will only dominate if the building is very tall). Columns carrying transfer trusses or similar massive loads may have high axial loads, and 1% of the factored axial load should always be considered in such cases. For any member also acting as a restraint to a column, a force of 1% of the column load needs to be resisted. Although the tie forces do not need to be considered with other forces, the restraint force required by BS 5950 Clause 4.7.1.2[8.3] does need combining with other member forces.

8.6.3   Vertical tying

During partial collapse, vertical ties can work with horizontal ties to share floor loads among all the floors and so they generally provide an additional level of robustness. Vertical tying is a formal requirement for Class 2B buildings but can normally be accommodated in steel structures because the columns are likely to be continuous. Caution is required in designing splices since it is only when columns have to act as emergency hangers that any significant tensions exist and it is that potential design load that governs for satisfying robustness. Fortunately, the minimum splice location details that are incorporated to aid assembly and erection will often equally suffice for code compliance; moreover, the demands of robustness ought in any case to be considered for the erection phase.

The demands of column splice tension capacity are covered in BS 5950[8.3] Clause 2.4.5.3. The numerical demands are that all column splices should be capable of resisting an axial tension equal to the largest factored vertical dead and imposed load reaction applied to the column at a single floor level located between that column splice and the next column splice down (or to the base). When applying this clause, it is the largest total reaction applied to the column at a floor level that should be used (i.e. the reactions from all the beams connected to the column at that floor level). The reactions at the floor level are calculated using the full imposed load and load factors.

SCI publication P212[8.14] has details of standard splices, and quotes axial tension capacities to simplify the design checks. Either bearing or non-bearing column splices can be used to satisfy vertical tying requirements. Non-bearing splices will generally have higher tension capacities because they inevitably require thicker cover plates and more bolts for normal design. Capacities are limited by bolt shear and adding additional bolts can easily increase capacities.

The vertical tying rules in BS 5950[8.3], Clauses 2.4.5.2 and 2.4.5.3, assume that columns are supported on foundations and continue vertically indefinitely, thus ignoring the cases of columns supported on transfer beams and also the column connections at the topmost level underneath transfer beams. To cover these cases, the top column connections below any transfer beam should be designed for the maximum tension that would occur if each column supporting the transfer beam was removed in turn. The base of columns on top of transfer beams should be designed for the standard vertical tying force.

Where the column is not continuous upwards (e.g. when supporting the topmost storey, which is probably light), the beam/column connections should be capable of taking the design load in both directions. Compliance with this guidance would appear to satisfy the requirements in AD-A[8.8] for Class 2B buildings. Class 2A buildings do not require vertical ties, so strictly would not need to comply, although it could be thought of as good practice if they did.

8.7 Bracing sytems

A principle of robustness is to provide redundancy or alternative load paths to cover for the event of unforeseen removal of part of the structure that would precipitate gross collapse. Many steel structures are designed as pin jointed frames and rely on bracing panels for lateral stability. Whilst numerically, the demands on the bracing system can be small and satisfied with a single bay, removal could cause considerable collapse. Such removal may even come about inadvertently during building conversion. Thus, it may be unwise to rely on a single bracing system, noting particularly that the capital costs of providing stability by bracing are normally small in relation to the remainder of the system. It would be reasonable to assume that a single robust concrete core would provide adequate resistance[8.15]. The purpose of the bracing systems provided should be highlighted on the drawings or in the Health and Safety File or more generally, the documents should make clear what global stability system has been presumed.

For Class 2B buildings, BS 5950-1[8.3] Clause 2.4.5.3 (d) requires that no substantial part of the structure (in each of two approximately orthogonal directions) is connected at only one point to the system resisting horizontal loading. Figure 8.2 shows one example. Resistance systems can be moment resisting joints, cantilever columns, shear walls, stair and lift cores, as well as triangulated bracing.

Figure 8.2 Multiple attachment points back to a bi-steel core with the remaining framing all tied in two directions

8.8   Floor units

Steel normally provides just the frame skeleton, the floors being either *in situ* concrete on metal decking (which may also act as a horizontal diaphragm) or precast units. There are provisions in BS 5950[8.3] regarding anchorage of the floor units down to the frame with the aim of preventing the floor falling through the frame if the steelwork is moved, or the floor units being uplifted as a result of accidental loading (e.g. explosion).

BS 5950-1[8.3] Clause 2.4.5.3 (e) requires that stairs, precast concrete or other heavy floor or roof units are effectively anchored in the direction of their span, either to each other over a support, or directly to their supports as recommended in BS 8110-1:1997[8.12] (see Chapter 7). The tie forces between floor units may be calculated from BS 8110-1 Clause 3.12.3 (see Chapter 7). Anchorage is only required in the direction of the unit span as the steel beams act as ties in the orthogonal direction.

Tying of the floor units to the beams will often be necessary for purposes other than reducing sensitivity to disproportionate collapse, such as to mobilise floor diaphragm action against wind loading.

The key reference 8.1 provides detailed guidance on anchorage of precast units across a range of support configurations. Further guidance on the use and design of precast units is also provided in SCI publication P287[8.16] and P351[8.17] and in Chapter 7. For Class 2B buildings, where precast units are the main structural support and the tying is provided by reinforcement within the topping, the units need to be connected to the steelwork ties i.e. rebar and mesh can be used for tying but that must be anchored or tied into the steelwork as well. Figure 8.3 shows typical details for Class 2A and Figure 8.4 for Class 2B.

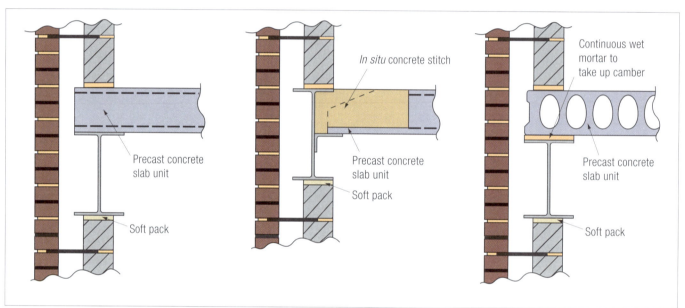

Figure 8.3   Anchorage of precast units to external steelwork (by friction and embedment). Class 1 and 2A

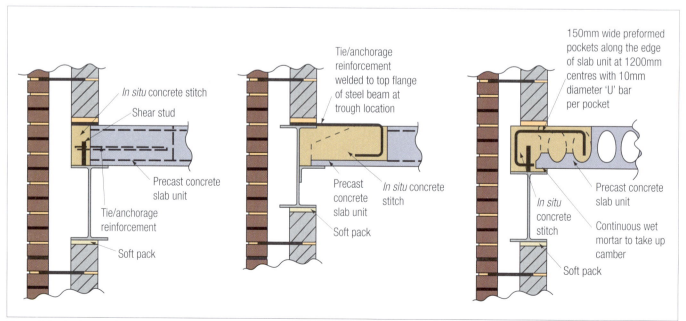

Figure 8.4   Anchorage of precast units to external steelwork (by positive ties). Class 2B

8.9 Notional removal of members

8.9.1 Introduction

When Clauses 2.4.5.3[8.3] (b), (c) or (d) are not satisfied, BS 5950 requires that notional removal of members or elements is considered as an alternative check on robustness. The principles are standard as described in Chapter 5 (including co-incident vertical loading) and involve notionally removing one element at a time to see if the structure will remain semi-intact without exhibiting disproportionate failure. If the notional removal of any element would result in the collapse of an area greater than the prescribed amount, then that element should be designed as a key element as recommended in BS 5950 Clause 2.4.5.4 (and below).

8.9.2 Notional removal of columns

Any element may be notionally removed. If a column is removed, the surrounding structure may be supported in a displaced state by relying on a combination of beam catenary action and hanging from the column above (just as in Chapter 5) (observing the requirement for sensible column splices and for the structure supporting the column as a hanger to be able to take the hanger load). From experience, the system capacity will most likely be determined by connection capacity. In checking this scenario, if any beams fail completely their debris load must be added to that of the floor below; dynamic impact need not be considered (see Chapter 4).

8.9.3 Notional removal of elements of the system for resisting horizontal forces

The highest risk of disproportionate collapse might be expected to occur if the lateral stabilising system is deleted and so for that reason, BS 5950[8.3] requires system redundancy (which can be satisfied via double connection points). Checking follows the standard process of removing one part of that system at a time and provided there is a back up system in place, there is a good chance that survival can be demonstrated, but there has to be a horizontal load path back to each system for both systems to be effective. As the only requirement is survival, the reduced stiffness of the single remaining bracing system should not matter so long as it is judged not so flexible as to permit generation of excessive de-stabilising effects.

If the lateral stability system relies on frame action utilizing moment resisting connections, then each element of the frame with a moment resisting joint is part of that system and should be notionally removed, one element at a time.

If the system for resisting horizontal forces is a concrete core, then each storey high segment of wall forming part of the core should be considered as an element of that system and notionally removed, one at a time. The length of loadbearing wall to be considered as one element, is as described in Chapter 5. As an alternative, it may well be possible to show that the core wall segment will function as a key element thus negating the need to consider its removal.

If the system for resisting horizontal forces is triangulated bracing, then each element of the bracing system should be notionally removed, one at a time. This includes the beam and column members forming part of the bracing truss.

8.10 Key elements

The requirements for key elements are standard i.e. they should be capable of resisting $34kN/m^2$ under the conditions described in Chapter 5 (which includes co-incidental vertical loading).

BS 5950[8.3] Clause 2.4.5.4 recommends that any other structural component that provides "lateral restraint vital to the stability" of a key element should also be designed as a key element and the same clause defines the manner in which load is to be applied.

Although transfer beams are not specifically required to be key elements, the beams and their column supports should be robust. And if any beam supports more than one column, it should probably be looked upon as a key element.

8.11 Resistance to extreme events

Continuous steel frames have inherent robustness via their ability to absorb energy through plastic deformation and even elastically analysed structures (with semi compact cross sections) may accept some redistribution. More information on this property can be found in Chapter 2 and in References 8.18. and 8.19.

8.12 Robustness of light steel frames

8.12.1 Introduction

BS 5950-5[8.10] provides detailed guidance on designing light steel frames for avoidance of disproportionate collapse. Guidance is also given in SCI publication P301[8.20]. Light steel multi-storey structures are generally constructed using a large number of regularly distributed structural elements, with a high degree of connectivity and structural integrity. In most applications, the provision of continuous ties between the components is straightforward because of the multiple inter-connections.

The following general rules for robustness for light steel frames are consistent with the principles used in BS 5950[8.3] for hot-rolled steel frames.

Every building frame (i.e. all building classes) should be effectively tied together at each principal floor and roof level. All wall studs should be anchored in two directions, approximately at right angles at each principal floor or roof which they support. This anchorage may be provided by either joists or tie members. Members such as steel decking, provided for other purposes, may be utilised as ties. When members are checked as ties, other loading may be ignored. Joists designed to carry the floor or roof loading will generally be suitable provided that their end connections are capable of resisting tension. Formulae for calculating the required tie force are given in BS 5950-5[8.10].

Class 2B should be designed to limit the effect of accidental removal of supports; this may be achieved by the standard methods of tying, notional removal or key element design.

8.12.2   Tying

The tying route is achieved by providing horizontal ties, vertical ties and a good distribution of vertical bracing throughout the building.

Horizontal tying should be arranged in continuous lines wherever practicable throughout each floor and roof level in two directions approximately at right angles. A tying member at the periphery of the building, for example at the head of a wall, should be connected back to the rest of the structure. If the vertical loads are resisted by a distributed assembly of closely spaced elements, the tying members should be similarly distributed to ensure that the entire assembly is effectively tied. The forces for anchoring the vertical elements at the periphery should be based on the spacing of the elements and taken as 1% of the factored vertical load in the element at that level. If the main structural elements are discrete columns, the horizontal ties anchoring the columns nearest to the edge of a floor or roof should be capable of resisting a factored tensile load, acting perpendicularly to the edge, equal to the greater of the load for an internal tie, or 1% of the factored vertical dead and imposed load in the column acting at that level.

All splices in primary vertical elements should be capable of resisting a tensile force of not less than two thirds of the factored design vertical dead and imposed load applied to the vertical element from the floor level below the splice. Unless the steel frame is fully continuous in at least one direction, the primary vertical loadbearing structural elements, whether discrete columns or panel walls, should be continuous at each beam-to-column/wall connection.

The system of bracing providing resistance to horizontal forces, whether discrete members or diaphragm panels, should be distributed throughout the building such that, in each of two directions approximately at right angles, no substantial portion of the building is connected to a means of resisting horizontal forces at only one point.

8.12.3   Notional removal

Where horizontal and vertical ties are not provided, the designer should check each storey to ensure that disproportionate collapse would not be precipitated by the notional removal, one at a time, of vertical loadbearing elements. If sufficient vertical bracing is not provided, a check should be made in each storey in turn to ensure that disproportionate collapse would not be precipitated by the notional removal, one at a time, of each element of the systems providing resistance to horizontal forces. For steel stud walls, the situation is considered to be similar to that of masonry and requires consideration of the optional removal of a wall panel. The length of an external stud wall that is considered to be removed is the length measured between vertical lateral supports. For internal steel stud walls that length is 2.25 times the storey height.

8.12.4   Key element design

Members or lengths of loadbearing wall that cause excessive areas of collapse when notionally removed should be designed as key elements. The design of key elements is similar to that described for hot-rolled steel frames.

Note that volumetric/modular construction differs from other forms of construction in that there is far more connectivity, such that a stack of modules may tolerate the notional removal of a whole module. In this case it is better to use a scenario-based approach. Guidance is provided in SCI publication P348[8.21] *Building design using modules*.

8.13   References

8.1   Way. A.J.G. *Guidance on meeting the robustness requirements in Approved Document A. SCI Publication P341*. Ascot: SCI, 2005

8.2   *BS EN 1993-1-1: 2005: Eurocode 3: Design of Steel Structures – Part 1-1: General rules and rules for buildings*. London: BSI, 2005

8.3   *BS 5950-1: 2000: Structural use of steelwork in buildings – Part 1: Code of practice for design – Rolled and welded sections*. London: BSI, 2001

8.4   *BS EN 1990: 2002: Eurocode: Basis of structural design*. London: BSI, 2002

8.5   *BS EN 1991-1-7: 2006: Eurocode 1: Actions on structures – Part 1-7: General actions – Accidental actions*. London: BSI, 2006 and *NA to BS EN 1991-1-7: 2006: UK National Annex to Eurocode 1 - Actions on structures – Part 1-7: Accidental actions*. London: BSI, 2008

8.6   *BS EN 1993: Eurocode 3: Design of steel structures* [Various parts]

8.7   *PD 6687: 2006: Background paper to UK National Annex to BS 1992-1*. London: BSI, 2006

8.8   Office of the Deputy Prime Minister. *The Building Regulations 2000. Approved Document A: Structure*. London: NBS, 2004

8.9   Steel Construction Institute. *Structural integrity of light gauge steel structures. Building Regulations Approved Document A (2000). SCI Advisory Note AD280*. Ascot: SCI, undated

8.10   *BS 5950-5: 1998: Structural use of steelwork in buildings – Part 5: Code of practice for design of cold formed thin gauge sections*. London: BSI, 1998

8.11 Steel Construction Institute. *Robustness rules for Slimdek. Document RT 1217.* Version 02. Ascot: SCI, 2008

8.12 *BS 8110-1: 1997: Structural use of concrete – Part 1: Code of practice for design and construction.* London: BSI, 1997

8.13 Steel Construction Institute. *Joints in steel construction: composite connections. SCI Publication P213.* Ascot: SCI, 1998

8.14 Steel Construction Institute and British Constructional Steelwork Association. *Joints in steel construction: simple connections. SCI Publication P212.* Ascot: SCI, 2002

8.15 Irwin, A.W. *The Design of shear wall buildings. CIRIA Report 102.* London: CIRIA, 1984

8.16 Hicks, S.J. and Lawson, R.M. *Design of composite beams using precast concrete slabs. SCI publication P287.* Ascot: SCI, 2003

8.17 Way, A.G.J. et al. *Precast concrete floors in steel framed buildings. SCI Publication P351.* Ascot: SCI, 2007

8.18 Hamburger, R. and Whittaker, A. 'Design of steel structures for blast-related progressive collapse resistance'. *Modern Steel Construction*, 44(3), March 2004, pp45-51

8.19 Yandzio, E. and Gough, M. *Protection of buildings against explosions. SCI Publication P244.* Ascot: SCI, 1999

8.20 Grubb, P.J. et al. *Building design using cold formed steel sections: light steel framing in residential construction. SCI Publication P301.* Ascot: SCI, 2001

8.21 Lawson, R.M. *Building design using modules. SCI publication P348.* Ascot: SCI, 2007

9 Timber: issues and solutions

9.1   Introduction

Reference should be made to Chapter 2 of this *Guide* for general guidance which is also applicable to buildings designed of timber.

BS 5268[9.3 – 9.6] and BS EN 1990[9.7] (EC0) both require that timber structures should be robustly constructed and various rules are provided to promote this objective. Essentially the same principles apply as for any other material. The first of these is that the inherent structural form should lend itself to assuring robustness; the second is that buildings should be designed for derived horizontal loads and thirdly all components need to be properly interconnected. The most common structural forms available for timber buildings are:
– trussed rafter roofs
– platform timber frame construction comprising loadbearing wall elements (e.g. studwork, structural insulated panels, cross laminated timber panels etc)
– post and beam frames
– portalised frames
and there are specific considerations for each type.

In the UK, trussed rafters are commonly used within hybrid structures; some longer span structures utilise glulam beams and platform timber frame construction is used for residential building up to seven storeys. Beam and post construction and glulam portal frames have been used for large span structures. Timber buildings commonly fall within Classes 2A and 2B.

Detailing of timber buildings requires attention to robustness at all stages of construction and at all material interfaces since the way many timber structures are constructed relies on considerable interaction between primary and secondary members. Although this potentially renders the structures vulnerable during assembly, once completed, the interconnection provides many opportunities for survival.

9.2   Robustness of timber structures

9.2.1   Introduction

Reference should be made to Chapters 4 and 5 of this *Guide* for general guidance which is also applicable to buildings designed of timber.

9.2.2   Roof structures e.g. trussed rafters

Tests[9.8] undertaken on a 10m span roof comprising 25° fink trusses have shown that trussed rafter roofs have inherent robustness. This was found to be the case both for suspended and plasterboard ceiling arrangements and also under unfavourable arrangements of wall plate end joints occurring above removed supporting walls. The four secondary components found to contribute most substantially to resisting collapse following wall removal were the wall plate, chevron bracing, top chord nodal bracing and even the tile battens with their interlocking tiling (full details can be found in Reference 9.8). (Although the reference tested trussed rafters, it is considered that bolted trusses would behave in a similar manner.)

Experience from the USA[9.9] indicates that most failures in timber trusses occur in individual members as a result of member shrinkage and connector slip yet do not result in the failure or collapse of the truss structure overall. Usually there is some readjustment among other framing members so that stresses are redistributed.

Part A3 requirements[9.10] make no direct reference to roofs but the British Standards for steelwork (BS 5950[9.11]) and masonry (BS 5628[9.12]) both incorporate a clause permitting the exclusion of horizontal ties in roofs of lightweight construction. Hence it may be argued that collapse of light timber roofs onto the floor below is unlikely to lead to floor failure and thence progressive collapse. On the other hand, in all timber construction, the ability of floors to sustain falling trusses may be less likely than in other materials and for this reason the design of timber floors below roofs should consider the possibilities of debris loading (see Section 5.10). Moreover, because timber elements are so light, they are more easily dislodged (for example roof trusses under wind suction), so fixings and restraint are especially important for assuring robustness, and caution is required in assessing the consequences of a roof failure.

9.2.3   Platform timber frame construction

Structures comprising loadbearing walls such as platform timber frames typically use materials easily fastened together and the benefits to robustness have been demonstrated by full scale testing as reported in Reference 9.13. However, despite the potential contribution of secondary items such as claddings and linings to the overall robustness of a structure, engineers should be relying on properly engineered load paths not potential load paths through finishes which may or may not be there depending on how they are fixed. Reliance on connections is only justified if the connection detailing rules of the relevant codes of practice are complied with.

Platform timber frame construction relies on a cellular plan form with all wall and floor components fixed to each other. Reliance is placed on the diaphragm action of the floors to transfer horizontal forces to a distributed layout of loadbearing walls. In turn, these walls combine vertical support with horizontal racking resistance. The format is inherently robust proven by many years of experience and through full-scale tests[9.13]. Robustness of the structural form exists as long as the following general principles are adopted:

– Provision of structural units that can be tied together. In particular, the fixing together of all intersecting walls should be in accordance with the minimum requirements given in Clause 4.9.2 of Reference 9.5.
– Ensuring that layouts and plan arrangements provide returns and intersecting walls and floors. Return walls should be provided at the ends of all loadbearing walls.

Consideration should be given to the layout of the loadbearing walls early in the design process to assist with providing a robust structure. Cellular layouts are best suited to multi-storey platform timber frames. Internal loadbearing walls may need to be strengthened to carry horizontal forces and where party walls separate the structure into separate units, the design has to ensure that horizontal forces can be taken by each unit. The opportunities for load transfer across party walls via structural ties are limited for reasons of maintaining acoustic performance[9.1].

Open plan layouts with no transverse structure are inappropriate for platform timber frames and additional structure is required for stability. The introduction of portal frame elements can provide solutions to open plan layouts, but attention is needed to maintain stiffness limits and connectivity of the framing types as well as controlling differential movement of different materials.

The timber codes BS 5268[9.3–9.6] and EC5[9.2] provide appropriate guidance for achieving strength and stiffness of the components that make up multi-storey timber frames and meeting their minimum standards is a pre-requisite for robustness. The approach for checking building stability is covered in BS 5268-6.1[9.5] for dwellings not exceeding seven storeys and a worked example can be found in Reference 9.14 (p39, topic: stability of platform timber frame). In particular, racking resistance, overturning and resistance to sliding should be checked.

In addition to designing a structure for stability by considering the equilibrium of forces and the strength and stiffness of elements, the stability or robustness of a building is also achieved through good practice detailing to ensure connectivity of items providing stiffness to the building. Examples of robust connections for timber frame floor and wall connections can be found in Reference 9.14 (p26, topic: robustness and floor to wall connectivity of platform timber frame).

Consideration must also be given to the asymmetric layout of elements intended to provide resistance to horizontal forces (e.g. racking walls). For structures or layouts that are unable to provide sufficient racking resistance, or have an unbalanced arrangement of racking resistance on plan, the use of discrete stiffening elements such as rigid frames, should be considered. The deflection limits for rigid frames should be appropriate for the structure and finishes but a limit of at least height/500 should be adopted.

Since BS 5268-6.1[9.5] is limited to 2.7m height panels, BS 5268-6.2[9.6] guidance should be used for panels above 2.7m (but limited to 4.8m) on all forms of building. The current limits suggested by BS 5268 parts 6.1 and 6.2 take a stiffness limit of the panel height divided by 300 as being acceptable to achieve deflection limits. The TF2000 tests[9.13] demonstrated that the actual stiffness of a timber frame building is significantly higher than the code limitations. Based on TF2000, there is evidence that the BS 5268 racking design principles are adequate for strength and stiffness for buildings up to eight storeys and that there is no reason for normal cellular platform frames to have additional deflection limits imposed.

9.2.4 Beam and post type frames e.g. glulam or engineered timber structures

Open-plan structures such as assembly halls, schools, retailing premises and hospitals may be constructed as beam-and-post timber frame structures. Provision of stability in the two orthogonal directions, temporary stability during construction and adequate loading allowances should be sufficient to ensure buildings are robust in both the permanent and temporary conditions.

For buildings where the failure of a single member (e.g. a column or principal truss) could cause a catastrophic collapse out of proportion to the element size, a relatively high degree of robustness should be provided. This can be achieved by designing the connections for the horizontal tie forces required by Annex A5 of EN 1991-1-7[9.15] (see Section 9.4.3). In addition, timber columns should be provided with fixed bases that are able to carry a moment and shear force equivalent to that resulting from the notional horizontal loads (see Section 9.5.1).

Single storey beam and post frames should have their roof structures adequately tied to the supporting structures in accordance with the guidance given in Section 9.4.2.

9.2.5 Timber portal frames

As for steel buildings, a special case arises for large span portals since failure of a single frame can results in a considerable collapsed area so potentially putting the structures into Class 2A or 2B.

There are similarities and differences between timber and steel portal frames. Steel portals have joints capable of plastic action which will differ from timber frame behaviour which should be assumed to behave elastically. The eaves joint on a timber portal frame must also be checked for combined bending and compression to ensure that lateral torsional instability of the timber sections does not occur. This is similar to the checks that would be undertaken for steel portal frames and is especially important where deep, slender sections are proposed. This design check should be carried out in accordance with Clause 6.3.3 of EC5[9.2].

Robustness should also be checked by assuring that there is no risk from longitudinal instability (by failure of wall bracing systems for example) and by assuring that the horizontal restraint to the portal base is adequate to resist both the forces from the permanent design load cases and the notional horizontal load case (see Section 9.5.1).

9.3   Design for the construction period

Caution is required in multi-storey construction where racking resistance and overall building stability depend on lining materials or vertical loads from the roof to contribute to stability. During construction, the building can be exposed to wind before such vertical loads are applied or before any plasterboard is fixed. Hence it is good practice to ensure that at least the lower half of the framing has adequate racking resistance from permanent bracing or sheathing materials alone (i.e. without contributions from plasterboard), since reliance on temporary bracing in large multi-storey construction has proven more complex than on low-rise projects.

For multi-storey frames, the construction period can be prolonged and so construction loading should be considered as a potentially destabilising effect. Significant vertical loads can be applied through storage of materials such as plasterboard packs and if stacked to one side these cause sway. Hence it is prudent to consider this sway possibility before plasterboard is fixed to stud walls, adding temporary restraints or additional members as appropriate to limit wall stud slenderness in the temporary condition.

9.4   Design for disproportionate collapse

9.4.1   Introduction

Reference should be made to Chapter 5 of this *Guide* for general guidance which is also applicable to buildings designed of timber.

The formal requirements of EN 1991-1-7[9.15] and BS 5268-2[9.3] are as follows in sections 9.4.2 and 9.4.3.

9.4.2   Framed structures e.g. post and beam

9.4.2.1   Class 2A buildings
For framed structures such as beam and post timber frames Clause A5.1 of EN 1991-1-7[9.15] (and Clause 1.6.3.3 of BS 5268-2[9.3]) require that horizontal ties should be provided and formulae (Expressions A1 and A2) enable the designer to calculate the required tie force for internal and peripheral ties.

9.4.2.2   Class 2B buildings
For framed structures such as beam and post timber frames, or structures which comprise a mix of beam and post and loadbearing walls, robustness should be provided by the provision of horizontal and vertical ties in accordance with Clause A5.1 and A6 of EN 1991-1-7[9.15] and Clause 1.6.3.3 and Clause 1.6.3.4 of BS 5268-2[9.3].

9.4.3   Loadbearing wall construction e.g. platform frame

9.4.3.1   Class 2A buildings
Because of the difficulty in providing horizontal ties in platform frame buildings, effective anchorage is normally the chosen approach. Clause A5.2 of EN 1991-1-7[9.15] and Option 1 of Clause 1.6.3.1 of BS 5268-2[9.3] states that appropriate robustness should be provided by adopting a cellular form of construction designed to facilitate interaction of all components including the provision of effective anchorage of suspended floors to loadbearing walls. This will generally be achieved in accordance with Clause 1.6.3.2 of BS 5268-2 and is satisfied by adopting industry-standard connection details between the floor edge and adjacent members. Further details of how to achieve effective anchorage are given in Section 9.6. Distinct horizontal and vertical ties are not required to be provided.

9.4.3.2   Class 2B buildings
For Class 2B buildings, robustness will be achieved by complying with the standard tying demands or by checking for notional removal of a loadbearing wall or by use of key elements (refer to Figure 5.1 and **Note b**). Because of the difficulty in providing horizontal and vertical ties in platform frame buildings, notional removal of loadbearing elements, one at a time in each storey of the building, in accordance with EN 1991-1-7 Clause A7 or Clause 1.6.3.5 of BS 5268-2[9.3] is the preferred approach. Further guidance on the design of timber platform frame buildings for Class 2B is given in Section 9.7.

9.4.3.3   Lightweight structures
Note that for both Class 2A and 2B structures, the UK National Annex to EN 1991-1-7[9.15] allows the minimum horizontal tie force to be limited in magnitude for lightweight building structures (whose primary structure is timber or cold formed thin gauge steel) to 15kN (Expression A1) and 7.5kN (Expression A2) for internal and peripheral building ties respectively in recognition of the reduced theoretical catenary tie force from a lightweight structure.

9.5   Notional horizontal loads and diaphragm action

9.5.1   Notional horizontal loads

Although not directly mentioned in EC5[9.2], clearly the general Eurocode approach of accounting for column out-of-verticality would also result in the generation of horizontal loads. Wind remains likely to be the source of dominant horizontal load, nevertheless, each storey should also have sufficient strength and stiffness in its own right to resist a horizontal, long-term force of 2.5% of the vertical load (defined as that percentage of the dead + imposed loads)[9.1].

9.5.2   Diaphragm action

BS 5268-2[9.3] requires that suitable bracing or diaphragm effect should be provided in planes parallel to the direction of the lateral forces acting on the whole structure. For all timber structures, but especially for multi-storey platform frames where robustness is provided by effective anchorage of suspended floors to loadbearing walls, it is necessary to check assumptions of diaphragm action to ensure the design is appropriate and specific checking is required at each level to assure that the transfer of horizontal forces is adequate. This should include the checking of all nailed interfaces to verify that appropriate in-plane and lateral shear forces can be transferred at each platform level. Reference 9.14 (P26, topic: robustness and floor to wall connectivity of platform timber frame) provides recommendations for good practice in fixings from platform floors to walls and for adequate mechanical capacity to transfer horizontal and vertical loads through the structure.

Special care should be taken where wall panels are not braced by horizontal diaphragms at regular intervals (e.g. full height external panels in stair cores) in which case wind posts may be required.

Where the action of an efficient ceiling diaphragm cannot be guaranteed (for example where a decoupled resilient bar type ceiling is provided), Figure M3 of BS 5268-2[9.3] indicates that additional edge blockings should be provided at floor perimeters where joists run parallel to the wall to ensure the lateral stability of the floor edge members (see Figure 9.1).

9.6 Application of ties to timber buildings

The tests on the TF2000 building at Cardington[9.13] demonstrated that correctly detailed timber frame structures should possess a degree of horizontal tying by the fixing densities provided between elements for overall building stability without the requirement for distinct horizontal ties.

Effective anchorage of suspended floors to loadbearing walls: to achieve the requirements of

effective anchorage, the approach for platform timber frame structures is to adopt good building practice by providing lateral restraint to walls and industry-approved anchorage details of suspended floors to walls. Additionally, the design process should involve checking the capacity of the component interfaces along the load path (e.g. panel rail to soleplate, soleplate to floor deck, floor joists to head binder and head binder to panel rail and so on) against horizontal forces. The designer should be providing robust connections at each and every junction as part of the normal design process (see Section 9.5.2).

A suspended floor can be considered effectively anchored if the connection between the floor and loadbearing walls has a minimum mechanical fixing specification in accordance with the details shown in either Figure M3 of BS 5268-2[9.3] (see Figure 9.1), UKTFA recommendations[9.14] (p 26, topic: robustness and floor to wall connectivity of platform timber frame) or BS 5628-1[9.12] Annex D for timber floors supported by loadbearing masonry. The UKTFA have also recommended minimum nailing densities in their Technical Bulletin 3[9.16] which will assure adequate anchorage of floors to walls.

In other building forms such as beam and post frames, where adoption of ties is possible, the tie forces can generally be accommodated by existing bolted connections in accordance with BS 5268-2[9.3] or EC5[9.2] as the partial safety factors for duration of loading and material strength for accidental actions generally make this load case less onerous than those for wind and vertical load.

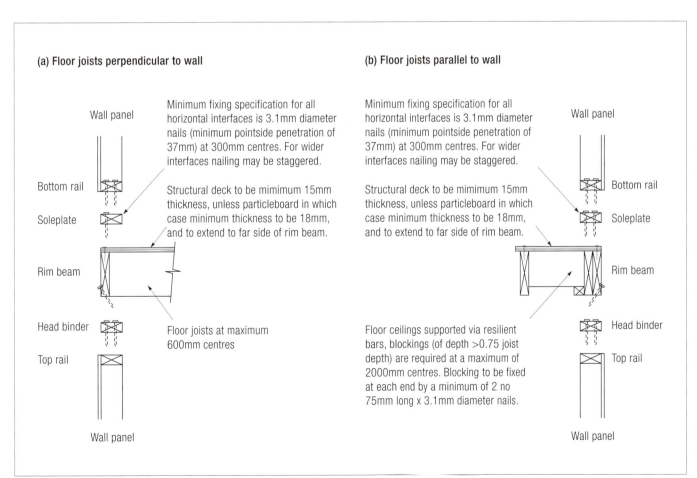

Figure 9.1 Extract from BS 5268-2:2002 Figure M3

9.7   Notional panel removal

In checking the robustness of timber framed buildings, engineers need to apply judgment based on the frame's likely 3-dimensional structural behaviour, backed, where appropriate, with a 2-dimensional structural assessment of discrete elements. The TF2000 full size test building[9.13] has shown this approach conservative yet appropriate for platform frame construction. The reference also provides guidance on the design process for Class 2B buildings where notional removal of loadbearing walls is part of the check.

The TF2000 test building[9.13] provided reassurance of the inherent robustness and availability of secondary load paths within platform frames. For example, sheathed walls with no openings can be regarded as deep beams with vertical shear taken in the panel to panel connections and tension taken out through the sheathing material in continuation with any timber framework across the panel junction e.g. via rim boards. Furthermore, the TF2000 tests[9.13] demonstrated that floors have reserve strength through the transverse spanning capacity of the floor sub deck and blockings when supported on walls parallel to the span or via upper walls acting as deep beams.

Unfortunately, the TF2000 tests[9.13] were specific to one building floor and panel shape and size. So the findings cannot be used to show universal compliance with the regulations and independent structural checks are required. It is possible to undertake such calculations to prove that wall panels can act as deep beams but often large or numerous openings exist, leading to a requirement for additional bridging members such as rim beams.

9.8   Rim beam method

A separate engineered timber rim beam, usually installed loose on site, can be used to span between points of vertical lateral restraint (say return walls) or key elements and act as a bridging member if loadbearing walls below are notionally removed.

Using rim beams allows joisted floor structures to be factory assembled as cassettes with a rim board used to connect the joist ends together for transportation. The rim beam can later function as a vertical load transfer element in the completed structure. The provision of a continuous rim beam ensures that structural continuity is achieved by providing vertical load transfer as a bridging element and horizontal continuity by providing a nailing density in accordance with the recommendations of Figure M3 of BS 5268-2 (see Figure 9.1) and UKTFA Technical Bulletin 3[9.16].

An example of the rim beam structural methodology is shown in Figures 9.2 and 9.3 and generic guidance can be found in Reference 9.13.

The rim beams need to react onto wall intersections and the wall returns must be of 1200mm minimum length (excluding framed openings). Such walls can be non-loadbearing in the conventional sense, but must still be capable of transferring loads down through the structure; the use of lightweight partitions built off floating floors is not acceptable. Any rim beams used are supported at the wall intersections by corner stud groups and to provide proper support, the beams must have full bearing onto the studs. To achieve this, the wall panels should be lapped in the opposite manner to the

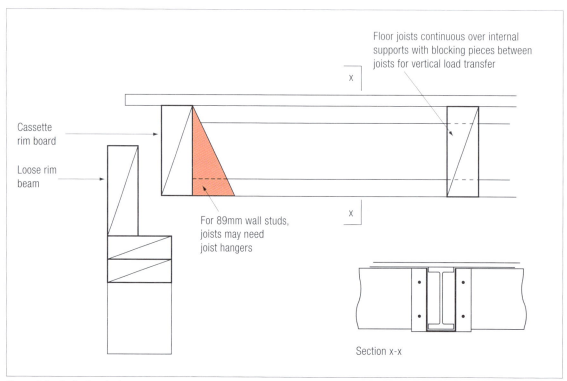

Figure 9.2   Indicative rim beam arrangement

rim beams. If no stud clusters are present below the rim beam bearing, hangers or fixings are to be provided off adjacent rim beams. Figures 9.4 – 9.10 indicate typical arrangements of rim beams and their supports.

Where the rim beams cannot be designed to span the required distance between return walls one, or more, intermediate posts will need to be provided and treated as a key element and designed accordingly (refer to Figure 9.9 and Section 9.9).

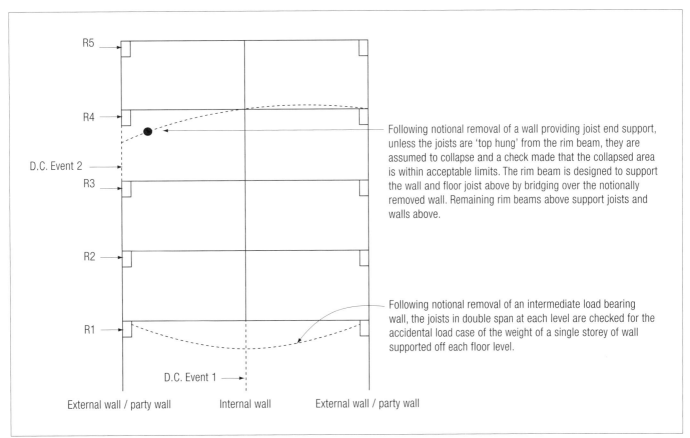

Following notional removal of a wall providing joist end support, unless the joists are 'top hung' from the rim beam, they are assumed to collapse and a check made that the collapsed area is within acceptable limits. The rim beam is designed to support the wall and floor joist above by bridging over the notionally removed wall. Remaining rim beams above support joists and walls above.

Following notional removal of an intermediate load bearing wall, the joists in double span at each level are checked for the accidental load case of the weight of a single storey of wall supported off each floor level.

Figure 9.3 Disproportionate collapse philosophy – rim beam method

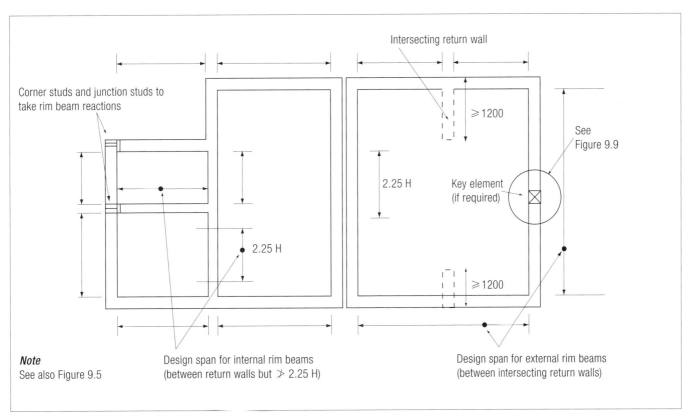

Note
See also Figure 9.5

Figure 9.4 Typical plan rim beam layout

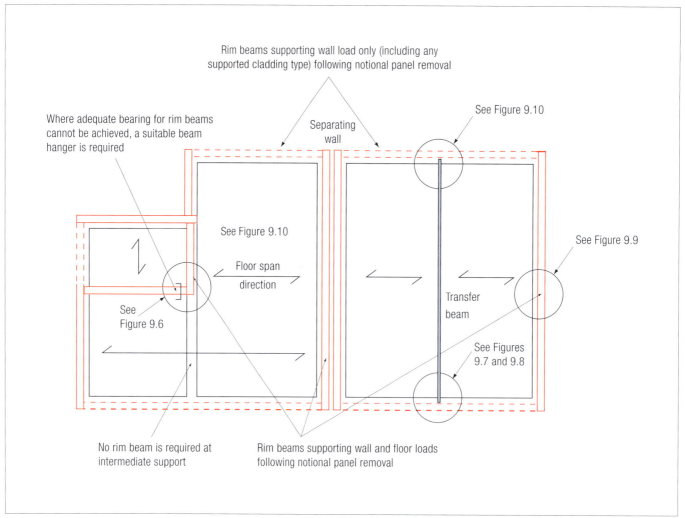

Figure 9.5   Plan view on a typical rim beam layout indicating the recommended arrangement of rim beam laps at wall sections

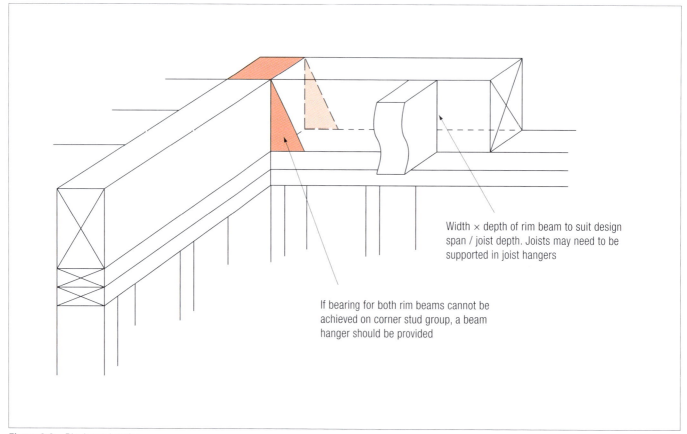

Figure 9.6   Rim beam junction

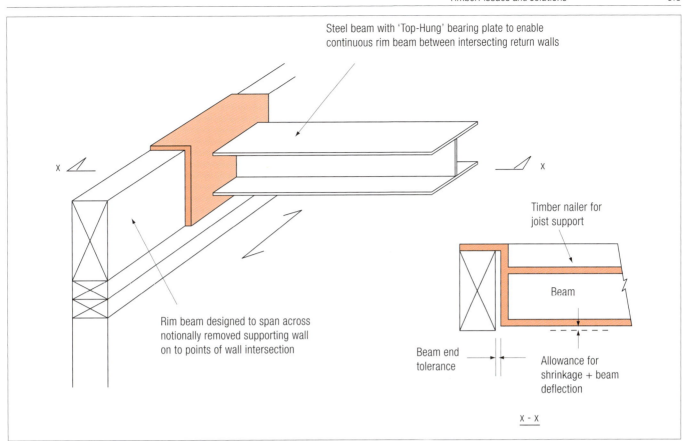

Figure 9.7   Transfer beam/rim beam junction option 1 – no key element post

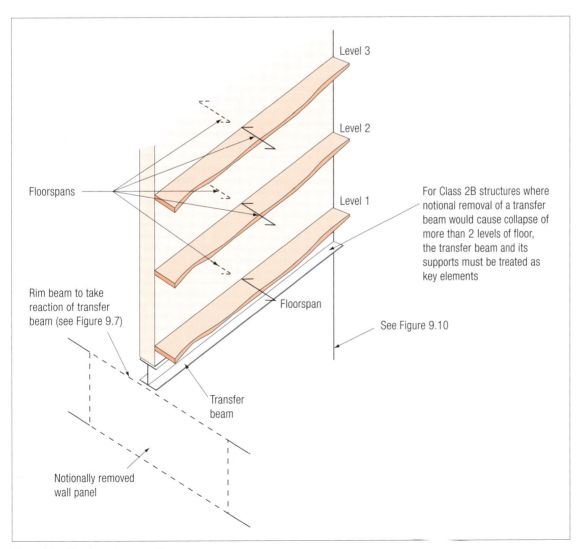

Figure 9.8   Key element posts and beams

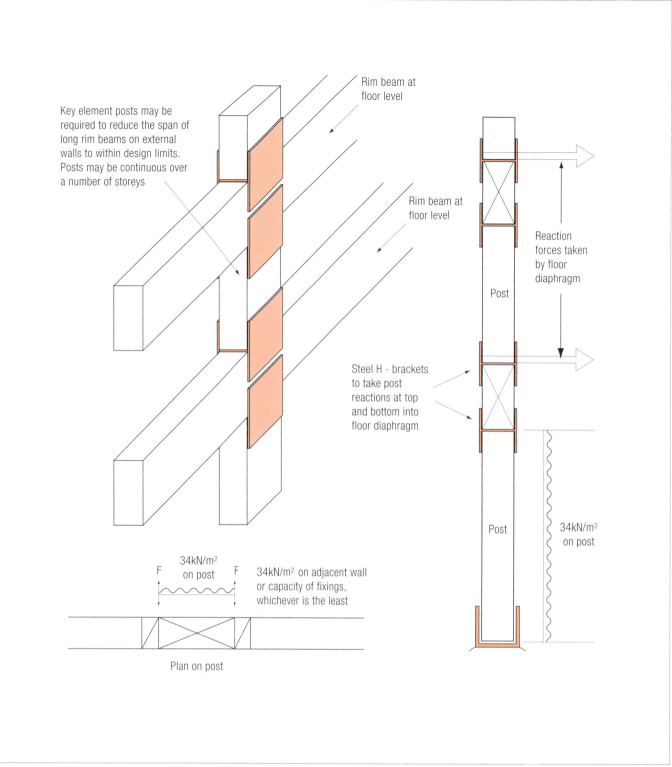

Figure 9.9   Key element posts acting in conjunction with rim beams

9.8.1   Assumptions for rim beam design

In designing rim beams, the standard assumptions about sequential removal (see Chapter 5) are applicable and the standard notional length of wall removed remains as:

- For external panels, the minimum length considered is 2.4m, with no maximum length. For internal walls, the maximum length of wall considered is 2.25H where H is the clear height of the panel between lateral supports (the top of the structural deck level below to the underside of the structural joist level above). Reference should be made to Clause 1.6.3.5 and Figure M2 of BS 5268-2[9.3] and Reference 9.1 which defines these lateral restraints in more detail.
- Rim beams and key element posts and their connections are designed to support the dead weight of the supported structure, one third of the imposed loads plus a single storey of wall panel with any supported claddings or linings.
- Rim beams must also provide horizontal tying action at all levels through the structure. The fixings presented for Class 2A buildings in Figure M3 of BS 5268-2 (see Figure 9.1) are the minimum required for robustness.

9.9 Rim beam and key element design principles

Key elements in timber buildings are designed as for other materials, for the notional 34kN/m².

The additional loads transferred by wall frame connections to a key element post (KEP) should be included in the loads applied to the KEP when checking its capacity and restraint at floor levels.

For design load cases appropriate for rim beam and KEP design EN 1990[9.7] Clause A1.3.2 and Table A1.3 and the supporting UK National Annex[9.7] (Clause NA 2.2.5) gives guidance on the accidental load combinations and psi (ψ) values to be used for the design of elements subject to accidental design situations (refer to Section 5.13). In addition, EC5[9.2] Table 3.1 gives k_{mod} for timber-based materials applicable to accidental actions. Values for short term action are appropriate for the residual loading effects following an accidental action.

BS 5268-2[9.3] Clause 1.6.3.7 gives equivalent guidance to that provided in EN 1990[9.7] for the residual structure design loads to be used in the design of key elements and rim beams. BS 5268 recognises the improbability of accident cases and the short term nature of the loading. Hence it allows a reduction of the co-existing load being carried (Clause 1.6.3.7[9.3]), and computation of residual capacity via enhanced factors for members (Clause 1.6.3.8[9.3]), and for fasteners (Clause 1.6.3.9[9.3]). A deflection limit of $L/30$ is applicable for timber elements supporting residual loads following an accidental event.

Transfer beams occur in timber structures supporting large floor areas. For general guidance on the design of these elements, reference should be made to Chapter 5. Prudence suggests checking these as key elements and this can be done by checking supporting members and by providing robust connections at their top and bottom (capable of resisting the stipulated 34kN/m²). Figure 9.8 indicates typical arrangements of key element posts and beams within loadbearing wall structures.

9.10 Other methods of designing against disproportionate collapse

Other possible methods of achieving support for walls and floors following the notional removal of wall panels exist, such as:
- The use of room-size floor cassettes with cassette edge boards acting as rim beams, bridging over removed wall panels. This is possible where small repeatable room sizes are present (e.g. hotel type accommodation).
- Loose floor construction with top-hung joists or joists supported in joist hangers from loose rim beams which bridge over removed wall panels.
- The use of wall panels designed as deep beams in lieu of rim beams – applicable where there are no openings in wall panels such as hotel bedroom dividing walls.

- Where joist span lengths are repetitive, using alternate double-spanning joists with selected walls designed as deep beams (a check should be made that overall building instability does not occur due to a lack of restraint at the back span of any cantilevered joists).
- Cantilevered joists supporting walls where the floor joists are designed to support the point load reaction from a single storey of wall panel plus any supported claddings following notional panel removal.

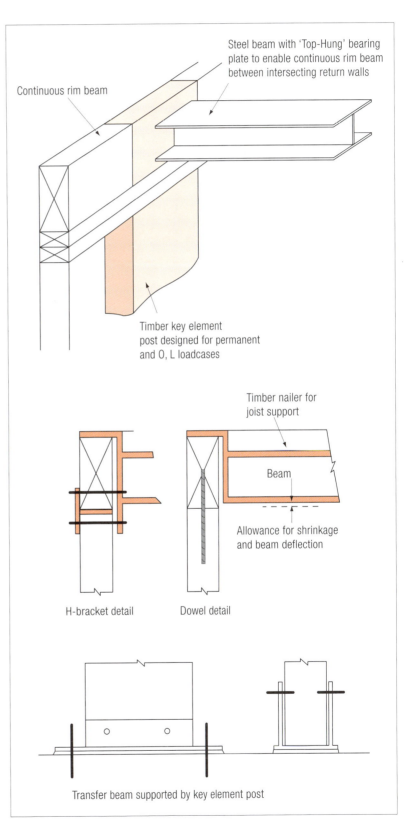

Figure 9.10   Key element posts – typical connection details

See Figure 9.11 for the structural methodology applicable to the cantilevered joist approach for satisfying disproportionate collapse.

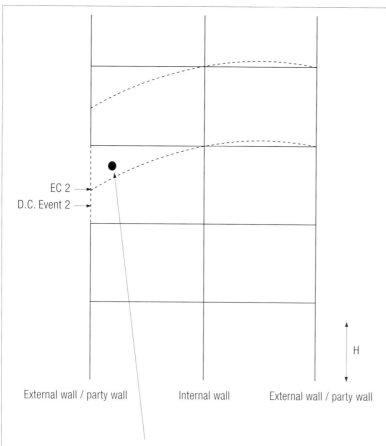

EC 2 →
D.C. Event 2 →

External wall / party wall   Internal wall   External wall / party wall

H

Following notional removal of a wall providing joist end span support, the joists are designed to cantilever and support a storey height of wall panel at the cantilever tip. Sufficient holding down resistance is required to prevent back span uplift (which can be reduced by staggering joist layouts)

Note
For the case of the removal of an internal loadbearing wall, the accidental load case for the joists is the same as for the rimbeam method (see Figure 9.3) with joists designed for an increased span length supporting the weight of the non-loadbearing partition at mid span.

Figure 9.11   Disproportionate collapse philosophy: Cantilevered joists method (notional removal of an external or compartment wall)

9.11 References

9.1   Grantham, R. and Enjily, V. *Multi-storey timber frame buildings: a design guide. BRE Report BR454*. London: BRE Bookshop, 2003

9.2   *BS EN 1995-1-1: 2004+A1: 2008: Eurocode 5: Design of timber structures – Part 1-1: General - Common rules and rules for buildings*. London: BSI, 2009

9.3   *BS 5268-2: 2002: Structural use of timber – Part 2: Code of practice for permissible stress design, materials and workmanship*. London: BSI, 2002

9.4   *BS 5268-3: 2006: Structural use of timber – Part 3: Code of practice for trussed rafter roofs*. London: BSI, 2006

9.5   *BS 5268-6.1: 1996: Structural use of timber - Part 6 – Code of practice for timber frame walls – Section 6.1: Dwellings not exceeding seven storeys*. London: BSI, 1996

9.6   *BS 5268-6.2: 2001: Structural use of timber - Part 6 – Code of practice for timber frame walls – Section 6.2: Buildings other than dwellings not exceeding seven storeys*. London: BSI, 2001

9.7   *BS EN 1990: 2002: Eurocode: Basis of structural design*. London: BSI, 2002

9.8   Marcroft, J.P. *Disproportionate collapse of timber structures. Part 1: Literature review; Part 2: Permissible stresses in fasteners and behaviour of timber connections under short duration loading; Part 3: Full-scale testing programme simulating accidental events on a trussed rafter roofed building. TRADA Technology Research Report RR 3/93*. High Wycombe: TRADA, 1993

9.9   Salgo, M.N. 'Examples of timber structure failures'. *ASCE Transactions*, 121, 1956, pp588-600

9.10 Office of the Deputy Prime Minister. *The Building Regulations 2000. Approved Document A: Structure*. London: NBS, 2004

9.11 *BS 5950-1: 2000: Structural use of steelwork in buildings – Part 1: Code of practice for design – Rolled and welded sections*. London: BSI, 2001

9.12 *BS 5628-1: 2005: Code of Practice for the use of masonry – Part 1: Structural use of unreinforced masonry*. London: BSI, 2005

9.13 Milner, M.W. et al. 'Verification of the robustness of a six-storey timber frame building'. *The Structural Engineer*, 76(16), 18 August 1998, pp307-312

9.14 UK Timber Frame Association. *Structural guidance for platform timber frame*. Available at: http://www.timber-frame. org/downloads/Structural_Guidance_for_Platform_Timber_ Frame.pdf [Accessed: 1 February 2010]

9.15 *BS EN 1991-1-7: 2006: Eurocode 1: Actions on structures – Part 1-7: General actions – Accidental actions*. London: BSI, 2006 and *NA to BS EN 1991-1-7: 2006: UK National Annex to Eurocode 1 – Actions on structures – Part 1-7: Accidental actions*. London: BSI, 2008

9.16 Milner, M. *Design guidance for disproportionate collapse. UK Timber Frame Association Technical Bulletin 3*. Available at: http://www.timber-frame.org/downloads/Disp_Collapse_ Bulletin_Final.pdf [Accessed: 1 February 2010]

10   Masonry: issues and solutions

Key reference: Masonry design for disproportionate collapse requirements under Regulation A3 of the Building Regulations (England and Wales)[10.1]. The Structural Masonry Designer's Manual[10.2] provides useful advice and Eurocode 6[10.3] (EC6) and BS 5628-1[10.4] are key references.

10.1   Introduction

Loadbearing masonry structures in brick or block are widely used for low rise structures and to some extent for higher rise structures such that any Class up to Class 2B might be relevant. Masonry supports are widely used within hybrid structures where the horizontal elements are timber or precast concrete floors. Dislodgement of the walls increases the risk of precipitating collapse, so the floor survival strategies discussed in Chapter 5 can be considered. Most traditional masonry receives its stability by being buttressed and by being loadbearing. High compression provides for transverse shear/friction resistance and high compression permits walls to carry lateral bending utilising pre-compression as a substitute for tensile strength or by facilitating resistance by arching to a degree which can be surprisingly high (always providing the arch lateral forces can be resisted). Lateral bending resistance in the transverse direction can be boosted by incorporating bed joint reinforcement and this might be used as an emergency span system to justify wall survival.

Needless to say, non-loadbearing masonry is potentially vulnerable to imposed lateral loads. Even loadbearing masonry may be sensitive if eccentric vertical loads are added as a result of poor bearing details. A robustness strategy should render all walls resistant to lateral loading for general reasons of safety and to avoid the need to consider taking their weight into account as debris should they be assumed to fail. As many non-loadbearing walls are internal with no defined horizontal wind loading, care should be taken to comply with code rules about height to wall thickness and to justify any structural assumptions about end restraint and anchorage. Particular care is required in the construction phase when all walls are vulnerable to any lateral loading.

The traditional cellular plan form of masonry structures offers inherent robustness if all the vertical and horizontal elements are interconnected and if sensible traditional practice with regard to wall end returns is deployed. Hence, as with all materials, achieving robustness starts with provision of a robust layout. But modern demands for open space or for making openings can inadvertently erode that traditional benefit. Additionally, masonry structures generally lack inherent tensile capacity (and ductility), and so become vulnerable to significant collapse in the event of accidental wall removal (see Box 4.1 of the Copenhagen gas explosion in Section 4.3.2).

EN 1991-1-7[10.5] is the Eurocode which addresses robustness issues (see Chapters 4 and 5). It gives guidance for design for accidental damage using similar terminology and principles to those used in Building Regulations Approved Document A[10.6] (AD-A). The Eurocode for masonry design, EC6[10.3] refers to EN 1991-1-7 but does not give any guidance on how the requirements of AD-A and EN 1991-1-7 are to be achieved. Such guidance is given in BS 5628[10.4]. With the replacement of British Standards by Eurocodes, this guidance will be transferred into a non-contradictory, complementary information (NCCI) document, so that it will not be lost. Unfortunately, there is some confusion in the published versions of the AD-A about the provision of horizontal ties in Class 2B buildings (see Section 5.1). It is understood that the need for horizontal ties in all Class 2B buildings, except in timber frame buildings, will be reinstated in the AD-A. In BS 5628 and other masonry industry publications, only one of the options for Class 2B buildings requires horizontal ties. Table 10.1 gives the BS 5628 recommendations. The NCCI will provide the finalised guidance.

Practical masonry buildings can have varying numbers of storeys or basements though the latter can be partly excluded from the number of storeys considered for classification (see Chapter 4 and Reference 10.7). There are risks of inadvertently reducing inbuilt robustness with change of use and with the modern trend of adding storeys or basements to existing buildings. Such changes potentially change the building classification so requiring a change of detailing which can be hard to implement.

New build loadbearing masonry buildings falling within Classes 2A or 2B can be dealt with using the codified prescriptive rules and the standard rules about tying in horizontal and vertical directions apply as with all other materials. But for Class 2B buildings, the alternative of accepting localised damage or considering local element removal may be found more useful since the incorporation of vertical ties is often difficult (see Section 10.4). Theoretically, the route of designing key elements is available but is often impractical.

Table 10.1   Detailed accidental recommendations adapted from Table 11 from BS 5628-1

Building class	Recommendations	
Class 1	Provide robustness, interaction of components and containment of spread of damage in accordance with the guidance in BS 5628[10.4] Clause 16	
Class 2A	As for Class 1, and additionally provide effective anchorage of all suspended floors to walls or effective horizontal ties in accordance with 33.4 and Table 12 of BS 5628	
Class 2B	As for Class 1, and additionally:	
	Option (1) Provide (other than key elements), supporting columns, beams or slabs supporting one or more columns or prove a loadbearing wall, or loadbearing walls removable, one at a time, without causing collapse	**Option (2)** Provide effective horizontal ties in accordance with 33.3 and Table 13, and vertical ties in accordance with 33.5 and Table 12 of BS 5628

10.2   Code requirements

As a generality, and noting that many masonry elements occur within hybrid structures, the advice that one engineer should be responsible for overall stability and robustness should be heeded. The defining code, BS 5628-1[10.4], reminds designers of this in Clause 16.1.

BS 5628[10.4] considers the sensitivity of masonry structures to overall instability. It prohibits using unbraced structural forms to resist lateral loading, e.g. pure cross wall structures that rely only upon wall flexural strength. The code states:

> "The design recommendations … assume that all the lateral forces acting on the whole structure are resisted by walls in planes parallel to these forces, or by suitable bracing."

and

> "To ensure a robust and stable design it will be necessary to consider the layout of structure on plan, returns at the ends of walls, interaction between intersection walls and the interaction between masonry walls and the other parts of the structure."

Within BS 5628[10.4] (Clause 16.1), overall robustness is addressed by requiring that the whole masonry structure (or any part of it up to roof level) is designed for a minimum overturning force whose notional minimum horizontal force value is 1.5% of the total characteristic dead (permanent) load above the level being considered.

10.3   Class 1 and Class 2A

The majority of loadbearing masonry buildings built in the UK are houses and hence at low risk of disproportionate collapse. Houses fall within consequence Class 1 of Eurocode 1[10.5] and Class 1 of AD-A[10.6]. Hence, provided they are designed and constructed in accordance with good practice (including carrying wind loads) and with basic stability rules, no further consideration over accidental actions need be given. Nevertheless, it may well be the case that walls not carrying vertical loads (such as flank walls in houses) require tying into the house floors or roof plane to give them stability and robustness as part of normal good practice.

The main point for Class 2A buildings is that all floor elements must be anchored to masonry walls so as to form effective horizontal ties in a similar manner to reinforced concrete structures and this can normally be achieved using routinely incorporated details. Appendix C of BS 5628[10.4] shows suitable anchorage details for various connections between masonry, timber and concrete elements: these are considered to have withstood the test of time in providing adequate robustness for masonry structures which fall within Class 2A. See also Chapter 7 and Figures 10.1 and 10.2 noting the difference between precast units spanning onto walls and units spanning parallel to walls.

Peripheral ties within the floor slab are primarily intended to deal with the loss of external walls, but may also be used to anchor the internal ties. Although in principle internal ties can be anchored to external masonry walls, it is doubtful if effective

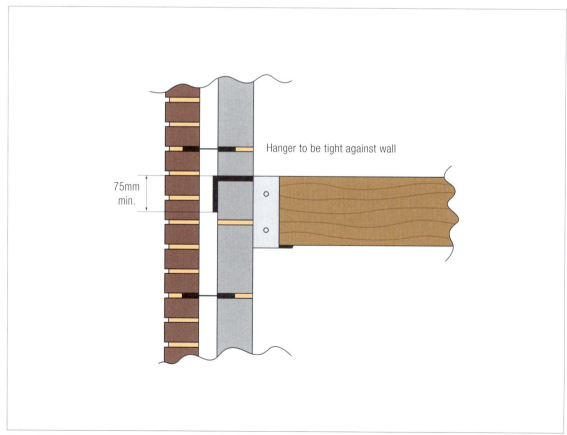

75mm min.

Hanger to be tight against wall

Figure 10.1   Floor anchorage to masonry walls: timber floor using nailed or bolted joints hangers acting as a tie

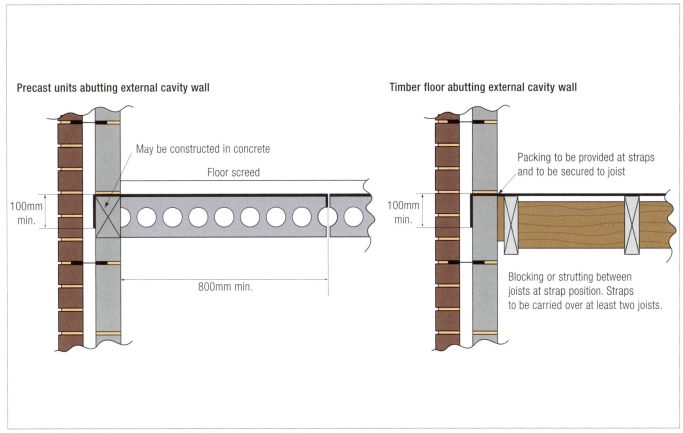

Figure 10.2 Floor anchorage to masonry walls: precast units and timber

anchorage can be achieved in practice without the concurrent use of vertical ties. Thus in masonry, it is usual to require the external walls and piers to be adequately connected to the floor construction to prevent their premature failure under outward pressure. This can be achieved by relying upon the shear strength of the connection, based on the type of masonry unit, mortar strength class and design vertical loading, or on its frictional resistance based on design vertical loading and appropriate coefficient of friction if the wall is loadbearing (but friction and shear capacity cannot be additive), or by utilising an *in situ* strip (see Chapter 7). Special joist hangers/ straps are available where timber floors are used.

Where ties are added, there are rules about their spacing and these are given in Table 10.2. Typically, horizontal ties are 30mm x 5mm section when designed to BS 5628[10.4] or BS 8103-2[10.8]. Ties should not be less than 600mm long and should cross over 2 joists (for example see Figure 10.2). However, AD-A[10.6] Section 2C, requires that ties are not less than 1.2m long and cross over 3 joists. For ties to be effective, the full load path has to be observed e.g. beam adequately connected to pad stone and pad stone adequately connected to wall.

Table 10.2   Spacing of ties (from BS 5628-1)

Construction	Spacing of horizontal restraint ties		
	Up to 3 storeys	4 storeys	5 storeys
Houses of Class 1 and 2A	2.00m	1.25m	1.25m
Other Buildings of Class 2A	1.25m	1.25m	Not applicable

10.4   Class 2B buildings: horizontal and vertical ties

For Class 2B buildings, the requirements are as set out in Table 10.1. Design will be based on the provision of effective horizontal and vertical ties or notional wall removal or use of key elements (though proving the adequacy of key elements is difficult in masonry, see Section 10.6).

Where sufficient vertical loading is available, shear or frictional resistance can be relied upon to provide the horizontal tie force. But the provision of the complementary vertical ties can be practically difficult and can only be achieved within masonry voids if the walling unit has a minimum thickness of 140mm, though other documents (e.g. EC6[10.3]) give 150mm (the minimum thickness allows room for ties and grouting). Special blocks are available which allow concreting or grouting up of the tie. Alternatively, discrete concrete columns (ties) can be formed in the wall space. The use of full storey height steel straps or steel sections (e.g. wind posts) as vertical ties is also possible, but the need to anchor these at floor levels can present difficulties. BS 5628[10.4] (Clause 33.5) has requirements for the minimum thickness and strength of walls incorporating ties and for their strength. Though, it is understood those minimum widths need not apply when using vertical steel straps or steel sections as ties since accommodation of internal bars is not a requirement.

Where provided, vertical ties should be distributed around the building. BS 5628[10.4] (Table 12) specifies maximum spacings and minimum forces to be resisted; as with horizontal ties, the forces to be resisted are related to the 34kN/m² accidental pressure. Where vertical ties are closely spaced, a minimum tie force is specified (BS 5628 Table 13) which corresponds approximately to the minimum percentage steel reinforcement required in plain concrete walls.

Whatever the form of vertical ties used, they must be taken down to the foundations and securely fixed there, or alternatively taken down to a level below which the vertical elements can be shown to function as key (protected) elements. Vertical ties must be effectively restrained horizontally at each floor level. BS 5628[10.4] Clause 33.5 also requires that walls containing ties must be constrained laterally to prevent movement and rotation. Timber floors are excluded from providing this function.

10.5   Class 2B buildings: notional element removal

In addition to their standard functions, vertical ties can also benefit masonry construction by providing a reaction to the vertical forces generated by the arching action of masonry walls, so enhancing the interaction between walls and floors when bridging an area of local damage, i.e. by enabling the wall/floor construction to act as a deep beam effectively with the slabs acting as flanges and the masonry between acting as a web.

If the notional element removal route is adopted, the approach is as set out in Chapter 5. The co-existing loads are to be taken as the design dead, imposed and, where appropriate, wind loads, using the reduced partial safety factors which are defined in BS 5628[10.4] (Clauses 18(d) and 3.3) or the accidental load case in the Eurocode[10.5].

10.6   Key elements

Within loadbearing masonry, only elements with substantial vertical loads will be capable of performing as key elements (i.e. taking the standard 34 kN/m²) because substantial pre-compression is required to generate vertical arching resistance. As a consequence, key elements will rarely be available as a practical design option (and normally only in tall buildings which will fall into Class 2B). Horizontal arching is also possible but often precluded where there are periodic vertical movement joints. Where walls or columns are investigated, a suitable design method is that using a failure model based on a three pin arch. To realise the arch model, adequate abutments must be present and the lateral thrust is as shown in Figure 10.3. (From simple equilibrium the value of q to fail the panel is $8nt/h^2$. Introducing a load factor of 1.05 reduces this to $7.6nt/h^2$ where n is the credible vertical reaction based on dead load and reduced live load.)

Within the arch, resistance to the horizontal reactions generated at the top and bottom can be provided either by the shear capacity between the masonry and concrete slabs or by friction. In either case, if it is assumed that both wall leaves form the contact surfaces, then that can only be justified if the outer leaf is adequately tied to the inner leaf.

When carrying out calculations, the vertical load is considered as the load available to resist the arch thrust and logic suggests that only permanent actions should be accounted for in generating the resistance to wall arching under accidental action. Chapter 5 discusses the load cases given in EN 1990[10.9] for the accidental damage case. Clause 18(d) of BS 5628[10.4] implies that the combination $0.95G_k + 0.35Q_k$ should be used for vertical arching calculations. In other situations, other combinations of partial safety factors on loads will provide the most severe condition. When considering accidental damage, material safety factors are halved.

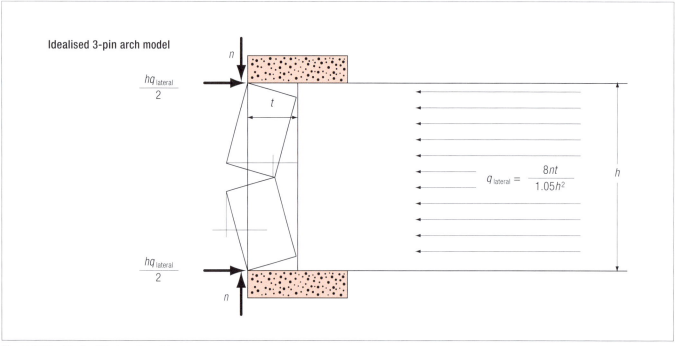

Figure 10.3   Idealised 3 pin arch model

10.7   Alterations to existing structures

Whilst the prescriptive rules applicable to buildings of Class 1, Class 2A and Class 2B may be suitable for some alterations, the difficulties associated with retrofitting ties, both vertical and horizontal can be formidable (Chapter 4 discusses rule implementation). In taller buildings, there may be a need to provide vertical tying below the level being altered. As an alternative, it may be possible to prove the section of wall or pier below is removable and then show that the masonry above (if sufficiently thick external walls or spandrel panels) can arch in-plane or be hung from the structure above.

Common alterations are the excavation of a basement beneath an existing structure, the removal of internal walls and the introduction of mezzanine floors. Where openings are created in existing cross walls or such walls are removed entirely to be replaced with beams, or when chimney breasts are removed, care must be taken not to endanger overall structural stability (degrade robustness). It is essential to consider the role of vertical loads in stabilising walls and great care is required during alterations if such load is removed (see Figure 10.4). The application of robustness requirements to altered buildings is discussed in Chapter 4.

Figure 10.4   Masonry collapse

10.8 References

10.1 Brick Development Association et al. *Masonry design for disproportionate collapse requirements under Regulation A3 of the Building Regulations (England and Wales).* Available at: http://www.brick.org.uk/_resources/Masonry%20Design%20for%20Disproportionate%20Collapse%20Requireents.pdf [Accessed: 1 February 2010]

10.2 Curtin, W.G. et al. *Structural masonry designers' manual.* 3rd ed. Oxford: Blackwell, 2006

10.3 *BS EN 1996-1-1: 2005: Eurocode 6: Design of masonry structures – Part 1-1: General rules for reinforced and unreinforced masonry structures.* London: BSI, 2005

10.4 *BS 5628-1: 2005: Code of practice for the use of masonry – Part 1: Structural use of unreinforced masonry.* London: BSI, 2005

10.5 *BS EN 1991-1-7: 2006: Eurocode 1: Actions on structures – Part 1-7: General actions – Accidental actions.* London: BSI, 2006 and *NA to BS EN 1991-1-7: 2006: UK National Annex to Eurocode 1 – Actions on structures – Part 1-7: Accidental actions.* London: BSI, 2008

10.6 Office of the Deputy Prime Minister. *The Building Regulations 2000. Approved Document A: Structure.* London: NBS, 2004

10.7 National House Building Council. *The Building Regulations 2004 edition – England and Wales: Requirement A3 – disproportionate collapse.* Available at: http://www.nhbc.co.uk/NHBCpublications/LiteratureLibrary/Technical/filedownload,23676,en.pd [Accessed: 1 February 2010]

10.8 *BS 8103-2: 2005: Structural design of low-rise buildings – Part 2: Code of practice for masonry walls for housing.* London: BSI, 2005

10.9 *BS EN 1990: 2002: Eurocode: Basis of structural design.* London: BSI, 2002

Bibliography

The topic of robustness and disproportionate collapse has occupied the international structural engineering community for many years with a noticeable first peak around forty years ago, as the implications of the Ronan Point collapse became evident. Since then, the technical literature has increased substantially, particularly so in the present decade in the wake of the World Trade Centre events. The following is a selective list – the 2010 Arup Security Report contains an extensive and up-to-date reference list – to help those interested in studying further this topic.

General

Arup Security Consulting. *Review of International Research on Structural Robustness and Disproportionate Collapse - Final Report*. Document Reference 125394-00/3.01/REP-001. Prepared on behalf of DCLG and CPNI, March 2010

Institution of Structural Engineers. *Safety in tall buildings and other buildings of large occupancy*. London: Institution of Structural Engineers, 2002

Institution of Structural Engineers. *Stability of buildings*. London: Institution of Structural Engineers, 1988

Best practice guides

Ellingwood, B.R. et al. *Best practices for reducing the potential for progressive collapse in buildings*. *NISTIR 7396*. Washington, DC: NIST, 2006. Available at: http://www.fire.nist.gov/bfrlpubs/build07/PDF/b07008.pdf [Accessed: 2 February 2010]

Institution of Structural Engineers. *Appraisal of existing structures*. 3rd ed. London: Institution of Structural Engineers [Due 2010]

Institution of Structural Engineers. *Manual for the design of steelwork building structures*. 3rd ed. London: Institution of Structural Engineers, 2008

Institution of Structural Engineers. *Standard method of detailing structural concrete: a manual for best practice*. 3rd ed. London: Institution of Structural Engineers, 2006

Textbooks/Monographs

Knoll, F. and Vogel, T. *Design for robustness. IABSE Structural Engineering Documents 11.* Zurich: IABSE, 2009

Paulay, T. and Priestley, M.J.N. *Seismic design of reinforced concrete and masonry buildings*. New York: Wiley, 1992

Starossek, U. *Progressive collapse of structures*. London: Thomas Telford, 2009

Technical papers

Alexander, S. 'New approach to disproportionate collapse'. *The Structural Engineer*, 82(23/24), 7 December 2004, pp14-18

Baker, J.W. et al. 'On the assessment of robustness'. *Structural Safety*, 30(3), 2008, pp253-267

Burland, J. 'Interaction between structural and geotechnical engineers'. *The Structural Engineer*, 84(8), 18 April 2006, pp29-37; Discussion, *The Structural Engineer Online*, 85(3), 6 February 2007, pp19-20

Byfield, M.P. and Paramasivam, S. 'Catenary action in steel-framed buildings'. *ICE Proceedings, Structures and Buildings*, 160(SB5), October 2007, pp247-257

Corley, G. 'Learning from disaster to prevent progressive collapse'. *ICE Proceedings, Civil Engineering*, 161(Special Issue 2), November 2008, pp41-48

Dusenberry, D.O. and Hamburger, R.O. 'Practical means for energy-based analyses of disproportionate collapse potential', *ASCE Journal of Performance of Constructed Facilities*, 20(4), November 2006, pp336-348

Ellingwood, B.R. and Leyendecker, E.V. 'Approaches for design against progressive collapse'. *ASCE Journal of the Structural Division*, 104(3), March 1978, pp413-423

Ellingwood, B.R. and Dusenberry, D.O. 'Building design for abnormal loads and progressive collapse'. *Computer-Aided Civil and Infrastructure Engineering*, 20(3), May 2005, pp194-205

Ellingwood, B.R. 'Mitigating risk from abnormal loads and progressive collapse'. *ASCE Journal of Performance of Constructed Facilities*, 20(4), November 2006, pp315-323

Gulvanessian, H. and Vrouwenvelder, T. 'Robustness and the eurocodes'. *Structural Engineering International*, 16(2), May 2006, pp167-171

Harding, G. 'Background paper on consequences of localised failure from an undefined cause'. *JCSS/IABSE Workshop on Robustness of Structures, BRE, Garston, 28-29 November 2005*. Available at: http://www.jcss.ethz.ch/events/WS_2005-11/Paper/Harding_Paper.pdf [Accessed: 2 February 2010]

Harding, G. and Carpenter, J. 'Disproportionate collapse of Class 3 buildings: the use of risk assessment'. *The Structural Engineer*, 87(15/16), 4 August 2009, pp29-34

Izzuddin, B.A. et al. 'Assessment of progressive collapse in multi-storey buildings'. *ICE Proceedings, Structures and Buildings*, 160(SB4), August 2007, pp197-205

Lawson, P.M. et al. 'Robustness of light steel frames and modular construction'. *ICE Proceedings, Structures and Buildings*, 161(SB1), February 2008, pp3-16

Maes, M.A. et al. 'Structural robustness in the light of risk and consequence analysis', *Structural Engineering International*, 16(2), May 2006, pp101-107

Marjanishvili, S.M. and Agnew, E. 'Comparison of various procedures for progressive collapse analysis'. *ASCE Journal of Performance of Constructed Facilities*, 20(4), November 2006, pp365-374

Menzies, J. 'Use of robustness concepts in practice'. *JCSS/IABSE Workshop on Robustness of Structures, BRE, Garston, 28-29 November 2005*. Available at http://www.jcss.ethz.ch/events/WS_2005-11/Paper/Menzies_Paper.pdf [Accessed: 2 February 2010]

Sasani, M. and Sagiroglu, S. 'Progressive collapse of reinforced concrete structures: a multi-hazard perspective'. *ACI Structural Journal*, 105(1), January-February 2008, pp96-103

Starossek, U. 'Disproportionate collapse: a pragmatic approach'. *ICE Proceedings, Structures and Buildings*, 160(SB6), December 2007, pp317-325

Sucuoglu, H. 'Resistance mechanisms in reinforced concrete frames subjected to column failure'. *ASCE Journal of Structural Engineering*, 120(3), March 1994, pp765-782

Appendix 1   London District Surveyors Association Risk Assessment Process

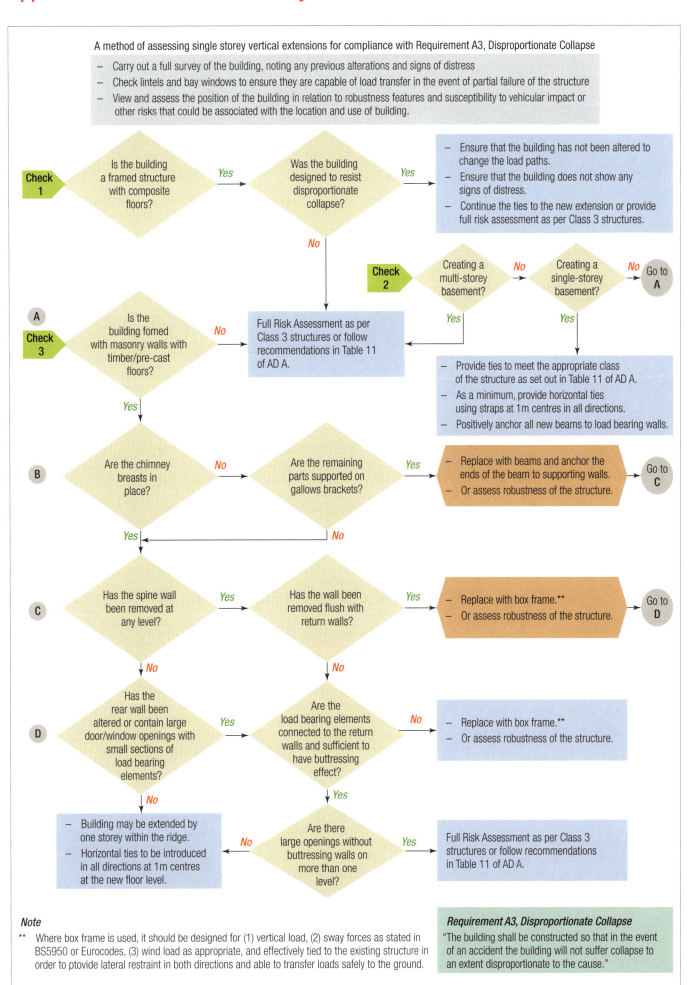

A method of assessing single storey vertical extensions for compliance with Requirement A3, Disproportionate Collapse

- Carry out a full survey of the building, noting any previous alterations and signs of distress
- Check lintels and bay windows to ensure they are capable of load transfer in the event of partial failure of the structure
- View and assess the position of the building in relation to robustness features and susceptibility to vehicular impact or other risks that could be associated with the location and use of building.

Check 1 — Is the building a framed structure with composite floors? → *Yes* → Was the building designed to resist disproportionate collapse? → *Yes* →
- Ensure that the building has not been altered to change the load paths.
- Ensure that the building does not show any signs of distress.
- Continue the ties to the new extension or provide full risk assessment as per Class 3 structures.

No (from "Was the building designed to resist disproportionate collapse?") ↓

Check 2 — Creating a multi-storey basement? → *No* → Creating a single-storey basement? → *No* → Go to A

Creating a multi-storey basement? *Yes* ↓ → Full Risk Assessment as per Class 3 structures or follow recommendations in Table 11 of AD A.

Creating a single-storey basement? *Yes* ↓ →
- Provide ties to meet the appropriate class of the structure as set out in Table 11 of AD A.
- As a minimum, provide horizontal ties using straps at 1m centres in all directions.
- Positively anchor all new beams to load bearing walls.

A — **Check 3** — Is the building fomed with masonry walls with timber/pre-cast floors? → *No* → Full Risk Assessment as per Class 3 structures or follow recommendations in Table 11 of AD A.

Yes ↓

B — Are the chimney breasts in place? → *No* → Are the remaining parts supported on gallows brackets? → *Yes* →
- Replace with beams and anchor the ends of the beam to supporting walls.
- Or assess robustness of the structure.
→ Go to C

Are the remaining parts supported on gallows brackets? → *No* ↓

Are the chimney breasts in place? → *Yes* ↓

C — Has the spine wall been removed at any level? → *Yes* → Has the wall been removed flush with return walls? → *Yes* →
- Replace with box frame.**
- Or assess robustness of the structure.
→ Go to D

Has the spine wall been removed at any level? → *No* ↓

Has the wall been removed flush with return walls? → *No* ↓

D — Has the rear wall been altered or contain large door/window openings with small sections of load bearing elements? → *Yes* → Are the load bearing elements connected to the return walls and sufficient to have buttressing effect? → *No* →
- Replace with box frame.**
- Or assess robustness of the structure.

Has the rear wall been altered or contain large door/window openings with small sections of load bearing elements? → *No* ↓
- Building may be extended by one storey within the ridge.
- Horizontal ties to be introduced in all directions at 1m centres at the new floor level.

Are the load bearing elements connected to the return walls and sufficient to have buttressing effect? → *Yes* ↓

Are there large openings without buttressing walls on more than one level? → *No* → (to "Building may be extended...")

Are there large openings without buttressing walls on more than one level? → *Yes* → Full Risk Assessment as per Class 3 structures or follow recommendations in Table 11 of AD A.

Note

** Where box frame is used, it should be designed for (1) vertical load, (2) sway forces as stated in BS5950 or Eurocodes, (3) wind load as appropriate, and effectively tied to the existing structure in order to ptovide lateral restraint in both directions and able to transfer loads safely to the ground.

Requirement A3, Disproportionate Collapse

"The building shall be constructed so that in the event of an accident the building will not suffer collapse to an extent disproportionate to the cause."